HIGHWAY TRAFFIC ANALYSIS AND DESIGN

HIGHWAY TRAFFIC ANALYSIS AND DESIGN

R.J. Salter

Reader in Civil Engineering
University of Bradford

SECOND EDITION

MACMILLAN

First edition 1974
Revised edition 1976
Reprinted 1978, 1980 (with amendments)
First paperback edition (with further amendments) 1983
Reprinted 1985

Published by
MACMILLAN EDUCATION LTD
Houndmills, Basingstoke, Hampshire RG21 2XS
and London
Companies and representatives
throughout the world

Typeset and Illustrated by TecSet Ltd, Wallington, Surrey
Printed in Hong Kong

0-333-48339-1

Contents

Preface

More than two decades have passed since transport and highway traffic engineering became recognised as an academic subject in institutions of higher education in the United Kingdom. During this period considerable developments in land use, transport modelling and the use of computer models of highway traffic flow have occurred. Considerable interest has also developed in the economic, social and environmental costs of highway and other transport systems.

The second edition of this book incorporates recent work on the fundamental principles of land use and transport planning methods and the subsequent appraisal of proposals. It reviews the analytical and practical aspects of highway traffic flow with sections discussing traffic noise generation, air pollution and the principles of road traffic restraint and road pricing.

Highway intersections are considered in detail with sections covering priority junctions, roundabouts, grade-separated junctions and interchanges. Junction control by traffic signal is covered in a comprehensive manner.

For the many engineers and planners engaged in transport work, but without the benefits of formal tuition, many sections of the book contain question and answer material so that the reader may test his comprehension of the subject matter by reference to the answers supplied.

The author would like to express his thanks to the many colleagues and students who have made valuable comments on the contents of the first and subsequently revised first editions. This second edition contains many of the suggestions they have made.

The author would like to express his appreciation to the following bodies for permission to reproduce their copyright material: Bedfordshire County Council, the Department of the Environment, the Building Research Station, the Eno Foundation, Freeman Fox and Associates, the Greater London Council, Her Majesty's Stationery Office, the Institution of Civil Engineers, the Institution of Highway Engineers, National Research Council, Royal Borough of New Windsor, Scott Wilson Kirkpatrick and Partners, Traffic Engineering and Control, the Transport and Road Research Laboratory, and Wilbur Smith and Associates.

PART 1
TRAFFIC ANALYSIS AND PREDICTION

1

Introduction to the transportation planning process

Large urban areas have in the past frequently suffered from transportation congestion. It has been recorded that in the first century vehicular traffic, except for chariots and official vehicles, was prohibited from entering Rome during the hours of daylight. While congestion has existed in urban areas the predominantly pedestrian mode of transport prevented the problem from becoming too serious until the new forms of individual transport of the twentieth century began to demand greater highway capacity.

Changes in transport mode frequently produce changes in land-use patterns; for example, the introduction of frequent and rapid rail services in the outer suburbs of London resulted in considerable residential development in the areas adjacent to local stations. More recently the availability of private transport has led to the growth of housing development which cannot economically be served by public transport.

In areas of older development however the time scale for urban renewal is so much slower than that which has been recently experienced for changes in the transport mode that the greatest difficulty is being experienced in accommodating the private motor car. Before the early 1950s it was generally believed that the solution to the transportation problem lay in determining highway traffic volumes and then applying a growth factor to ascertain the future traffic demands.

Many of the early transportation studies carried out in the United States during this period saw the problem as being basically one of providing sufficient highway capacity and were concerned almost exclusively with highway transport.

During the early 1950s however it was realised that there was a fundamental connection between traffic needs and land-use activity. It led to the study of the transportation requirements of differing land uses as the cause of the problem rather than the study of the existing traffic flows. The late 1950s and early 1960s saw the commencement of many land use/transportation surveys in the United Kingdom and the era of transportation planning methodology could have been said to have commenced.

Because the planning of transportation facilities is only one aspect of the overall planning process which affects the quality of life in a developed society, the provision of transport facilities is dependent on the overall economic resources available. It is dependent on the value that is placed on such factors as environmental conditions; for some transport facilities are considered to detract from the quality of the environment and others can be considered to improve the environment. Land use and transport planning are also closely connected because the demand for travel facilities is a function of human land activity and conversely the provision of transport facilities has often stimulated land-use activity.

Because we are living in a society that is changing rapidly, and in which the rate of change appears to be increasing, it is important for some attempt to be made to develop economic, environmental, land use, population and transport planning policies. The fact that planning attempts in all these fields have not met with conspicuous success in the past decade should be taken as an attempt to improve the methodology rather than an indication that short term plans based on expediency or intuition should be employed.

Transportation studies may be carried out to determine the necessity or suitability of a variety of transport systems such as inter-city air-links, a new motorway or a combination of private and public transport modes such as is found in a large urbanised conurbation. The methodology of these surveys will vary in detail—but most transportation surveys that are based on land-use activity tend to be divisible into three major sub-divisions.

(i) The transportation survey, in which an attempt is made to take an inventory of the tripmaking pattern as it exists at the present time, together with details of the travel facilities available and the land-use activities and socio-economic factors that can be considered to influence travel.
(ii) The production of mathematical models, which attempt to explain the relationship between the observed travel pattern and the travel facilities, land-use activities and socio-economic factors obtained by the transportation survey.
(iii) The use of these mathematical models to predict future transportation needs and to evaluate alternative transportation plans. These three stages of the transport planning process are illustrated in figure 1.1, which shows the procedure used to estimate future travel in the Greater London Area.

In the first stage, details of the existing travel pattern together with information on land use and transport facilities are obtained for the area of the study. This area is bounded by an external cordon and so that the origins and destinations of trips within the area can be conveniently described, the study area is divided into traffic zones.

Details of the existing travel pattern are obtained by determining the origins and destinations of journeys, the mode of travel and the purpose of the journey. Most surveys obtain information on journeys that have origins in the survey area by a household interview method, which records details of the tripmaking of survey-area residents. In addition there will be some trips that have origins outside the external cordon and destinations within the cordon and others that have neither origin nor destination within the survey area but pass through the study area. Details of these trips will be obtained by interviewing tripmakers as they cross the cordon. Additional surveys will be necessary to obtain details of commercial vehicle trips originating in the survey area and in some circumstances trips made by means of taxis.

Information on transport facilities will include details of public transport journey times, the frequency of service, walking and waiting times. For the road network, details of traffic flows, journey speeds, the commercial vehicle content and vehicle occupancies are frequently necessary.

As land-use activity is the generator of tripmaking, details of land-use activity are required for each traffic zone. For industrial and commercial land use, floor space and employment statistics are necessary, while for residential areas, the density of development is frequently considered to be of considerable importance. At the same time, socio-economic details of the residents are obtained since many surveys have indicated a connection between tripmaking and income, and also between trip-making and social status.

In the second stage of the transportation process, models are developed which attempt to explain the connection between the tripmaking pattern and land-use activity. This is the most difficult aspect of the transportation planning process in that failure adequately to determine the factors involved in tripmaking will result in incorrect predictions of future transport needs.

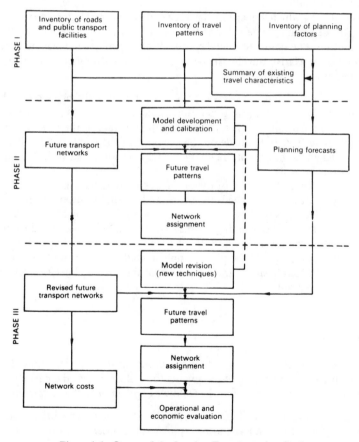

Figure 1.1   Stages of the London Transportation Study

While the decision to make a trip is a complex process based on the availability of destinations, the travel facilities, the cost of travel and the journey purpose, it is usual to divide model building into the following interconnected processes which are described in detail in the subsequent sections.

(a) Trip generation, which attempts to determine the connection between tripmaking and land-use factors noted in the planning inventory.

(b) Trip distribution, which determines the pattern of trips between the zones. It is usually postulated that the number of trips between zones is proportional to the size of the zones of origin and destination and inversely proportional to some measure of their spatial separation.

(c) Traffic assignment, which decides on which links of either the public transport or the highway network a trip will be made. On the basis of the travel cost, the decision is made on the alternative routes between the origin and the destination. Trips are usually assigned to the route of least cost, frequently measured by travel time. More recently a proportion of trips has been assigned to the alternative links between an origin and a destination in an attempt to produce a realistic simulation of the real life situation.

(d) Modal split, by which a decision is made as to which travel mode a tripmaker will use. This decision may be made at the trip generation stage or frequently after trip distribution. The observed modal split relationship may be explained in terms of car ownership, relative travel costs between the alternative modes or on the basis of restrained private vehicle use.

After these interrelated processes have been set up, the existing land-use characteristics are inserted into the trip generation procedure. The trips are distributed, assigned and a modal split decision at the appropriate stage of the process carried out. This allows checks to be made on the accuracy of the whole process because the flows assigned to the present day network may be compared with those actually observed on the network.

With confidence in the ability of the developed models established, it is then possible to forecast the travel needs of future land use and transport plans. The evaluation of the effects of these alternative proposals is the object of the third stage of the transportation planning process. Arriving at an optimum solution is however an intuitive process because the planning process can only predict the likely tripmaking which will arise if the plan is implemented. Alternative plans may be evaluated, on a limited basis, by the estimation of the costs and benefits which arise if the plan is carried out.

Problem

During the period 1950–60 an important reappraisal of thinking regarding the estimation of future transportation requirements took place. Select the correct description of this reappraisal from the following and amplify the statement:

(a) It was appreciated that the highway network should be designed to accommodate at least three times as many motor vehicles as are observed on the existing network.

(b) It was appreciated that there was a basic connection between land-use activity and transportation requirements.

(c) It was appreciated that the existing highway network in the centres of large urban areas would require extensive modification.

Solution

(a) While a considerable increase in the number of motor vehicles in use can be expected in the future it is not considered desirable simply to scale up the existing highway network. The demand for transport facilities will increase at differing rates in differing situations and not all of these demands are likely to be met by the provision of enlarged highway facilities.

(b) The realisation that the demand for transport facilities was a function of land use caused an important reappraisal in the estimation of future transport needs. Instead of observing the end product and applying some growth factor to estimate future values it became possible to obtain the traffic requirements of differing land-use plans. It changed transport planning from a vehicle count to a land-use-based operation.

(c) While it was appreciated that the modification of the highway network in the centres of large urban areas would be required in the future it is considered desirable to restrain the use of motor cars in many central areas on environmental grounds.

2

The transportation study area

The area of a country covered by a transportation study will depend upon the geographical distribution of the trip patterns which are of particular interest to the study organisers. The study area has been defined as the area within which trip patterns will be significantly affected by the implementation of any resulting transport proposals. It is important that sufficient thought be given to the boundary of the area because within the boundary the transport information is collected in considerable detail and the resulting cost of data collection will be largely related to the location of the boundary.

Some surveys are concerned with individual towns of small size and in these cases the area may cover only the developed land together with land which may become developed during the economic lifetime of the transport proposals which may result from the study.

The boundary of the study area within which trip information is collected and modelled in considerable detail is termed the external cordon.

It is not possible to confine transportation surveys to existing local government boundaries because tripmaking and transportation needs are common to a region. Most surveys are carried out on a regional basis; for example, a West Yorkshire transportation Study[3] covered an area between Knaresborough and Barnsley, Burnley and Goole. The London Travel Survey area contains almost 17 per cent of the population of Great Britain bounded by St. Albans, Slough, Gravesend and Redhill.

The survey area within which travel is to be studied in detail is bounded by an external cordon. The position of the external cordon is fixed so that as far as possible all developed areas which influence travel patterns are included together with areas which are likely to be developed within the forecasting period of the study. Within the external cordon, trip information is collected in considerable detail and so the cost of the survey is largely related to the cordon location.

To permit the aggregation of data, the survey area is divided into internal traffic zones; the remainder of the country external to the cordon is also divided into considerably larger external zones. The Department of the Environment has divided the country into Standard Regions and this is useful for the external traffic zones.

Zones are selected after consideration of the following factors.

1. They should be compatible with past or projected studies of the region or adjacent region.

2. They should permit the summarising of land use and tripmaking data.

3. They should allow the convenient assignment of trips, which are assumed to be generated at the centroids of traffic zones, to the transportation network. There is often conflict here between zones which are suitable for the highway network and zones suitable for the public transport network.

4. While greater accuracy is possible with smaller traffic zones they should not be so numerous as to make subsequent data processing difficult. In the Los Angeles Regional Transportation Survey there were 2346 'minor' traffic zones which were combined into 274 'major' zones[1]. The London Travel Survey covered an area of 941 square miles and contained nearly 3 million households. There were 933 zones, 186 districts and 10 sectors[2].

5. The size of the zones will be governed by the detail which is required in the modelling process and in the resulting transport proposals. In an early study which was mainly concerned with strategic highway needs in the Yorkshire area[3] the whole of the then cities of Bradford and Leeds were divided into 3 and 4 traffic zones. In contrast the more recent transportation study of West Yorkshire carried out by WYTCONSULT[4], which was concerned with determining transport needs in great detail over the short, medium and long term, employed 1816 internal zones and 223 external zones with Bradford and Leeds being divided into 216 and 324 zones respectively.

6. Because of the increasing use of data from the Registrar General's Census of Population the zones should be selected after consideration of enumeration, district boundaries and the grouping of local planning authority data.

7. As data is frequently aggregated by zone, the latter should be areas of predominantly similar land use activity in which similar rates of future growth are anticipated.

When the traffic zones are being determined, it is important to plan them with thought for the provision of one or more screenlines. A screenline is a line which

——————— Zone boundaries

━━━━━━ Sector boundaries

Figure 2.1 Internal traffic zones

divides the area within the external cordon and it is chosen so that the trips cross-ing it at the present day can be easily measured. For this reason a natural barrier to communication should, if possible, be used so as to make it possible to count the traffic movements across the screenline. It should be so positioned that it is unusual for trips to cross the screenline more than once. To avoid the complex traffic move-ments which take place in the central sector a screenline is frequently placed to intercept all the movements into the central area.

An illustration of a hierarchy of traffic zones is given in figures 2.1, 2.2, 2.3 and 2.4 showing the arrangements used in the Bedford-Kempston Transportation Survey[5]. Figure 2.1 shows the internal zones varying in area from the small intens-ively developed Central Area zones to the larger zones in the outer area of Bedford. These internal zones are grouped into a central area sector and eight radial sectors. The sector numbers 0-8 inclusive are the first digit in the internal zone code numbers.

Within the external cordon, two screenlines were set up for checking purposes and this is shown in figure 2.2. Also shown are the screenline counting stations and also the external cordon stations.

Outside the external cordon but immediately adjacent to the survey area the external zones are comparatively small, generally corresponding to local authority areas or combinations of areas. These zones are shown as intermediate external zones in figure 2.3. The remainder of the country is divided into national external zones as shown in figure 2.4. All the external zones are numbered in accordance with the Department of the Environment standard classification.

Department of Transport National Zoning System

During the mid-1970s the Department of Transport commenced the development of a Regional Highway Traffic Model in order to establish a consistent and rational basis for the prediction of traffic flows on inter-urban highways.

As part of this project a nationally consistent system of zones was defined that was compatible with local authority areas and that could be used for the planning of road networks. A two-stage system was evolved that was suitable both for major studies in which many of the trips being modelled would have origins and destina-tions in widely spaced regions of Britain, and also for more local schemes.

In land use, transport-planning zones are used not only for the assignment of vehicle movements but also as a framework for the collection of the land-use data that is vital for the modelling process. Because much of the land-use data such as population, household numbers and employment relates to local authority bound-aries, the zoning system developed used County and District Council boundaries.

The system of zones defined in the Regional Highway Traffic Model comprised County zones, District zones, Strategic zones (formerly known as National zones), Local (RHTM) zones (formerly known as Regional zones) and scheme zones.

The zones at each level are formed by aggregating zones at the next lower level with the County and District zones that correspond to the administrative area formed after the local reorganisation in 1974. For national road-planning purposes the smallest traffic zone is the local zone.

The final outcome of the division of Britain into traffic zones was that there were 78 County zones, 447 District zones, 1190 Strategic zones and 3613 Local zones. Average Local zone population is 13 500 in England with individual zones ranging from 50 to 50 000, higher figures being found in parts of London.

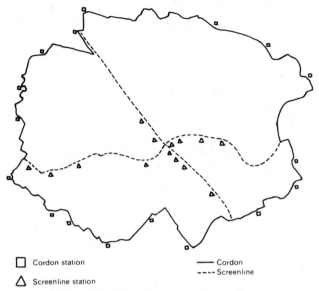

Cordon station

Screenline station

Cordon
Screenline

Figure 2.2   External cordon and screenlines

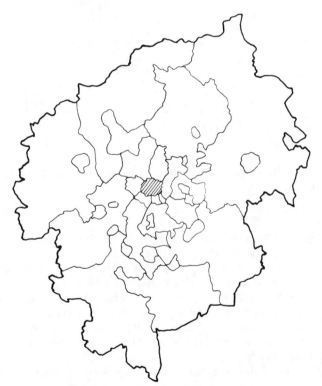

Figure 2.3   External traffic zones (intermediate)

Figure 2.4   External traffic zones (national)

The expected advantages of this system of zoning are given in the Traffic Appraisal Manual[6].

(a) It is a national framework of traffic zones consistent with local authority administrative areas.

(b) Zone boundary files are available with Ordnance Survey references for plotting at any level of aggregation or allocation of other data.

(c) It provides a link with previous sources of traffic data, in particular the previous Department of Transport and Planning National Traffic Model.

(d) Data can be cross-referenced to postcodes, which are linked to the zone system.

(e) If this system is used for future data collection it will ensure consistency.

(f) Land-use data forecasts are already available to this zoning layout.

(g) The zoning system allows zonal information to be related to an existing or proposed highway network.

(h) A single point location is available which represents the zone area at any zone level.

(i)   In the development of the Regional Highway Traffic Model standard computer software has been developed for traffic-modelling purposes using the zone numbering principles.

In a transportation study recently carried out for Baghdad by Scott Wilson Kirkpatrick and Partners the study area encompassed an area of 927 sq km and had an external cordon length of 154 km. The study area and the remainder of Iraq was divided into 301 internal traffic zones and 72 external zones. The internal zones were grouped into 13 sectors which were in turn sub-divided into 31 coarse zones. Screenlines used to determine the accuracy of the modelling procedure were drawn whenever possible along barriers to movement which included the River Tigris, the Army Canal and certain railway lines. The zoning system which was used was based upon criteria such as the use of logical catchments sufficient for satisfactory traffic modelling, compatibility with previous planning and census work and with existing district boundaries.

References

1. Transportation Association of Southern California, *Los Angeles Regional Transportation Study, Appendix to the Base Year Report*, Los Angeles (1969)
2. London County Council, *London Traffic Survey*, County Hall, London, Vol. 1 (1964)
3. Traffic Research Corporation, *West Yorkshire Transportation Study, Project Report* (1967)
4. B. V. Martin and P. N. Daly, The West Yorkshire Transportation Studies (1) The analytical approach, *Traff. Engng Control*, **19** (1978), 2, pp 536-40
5. Bedfordshire County Council, *Bedford-Kempston Transportation Survey, Phase 1* (1965)
6. Department of Transport, *Traffic Appraisal Manual*, London (1982)

3

The collection of existing travel data

Existing tripmaking or travel data may conveniently be classified into four groups according to the origin and the destination of the trip being considered. These four classes of movement are:

(a) trips which have an origin within the external cordon and a destination outside the external cordon;

(b) trips which have both an origin and a destination within the external cordon;

(c) trips which have an origin outside the external cordon and a destination within it;

(d) trips which have neither origin nor destination within the external cordon but which pass through it.

Details of these trips are obtained in differing ways. Trips of type (a) and (b) are usually recorded by means of a home interview survey while trips of type (c) and (d) are normally recorded by means of an origin–destination survey conducted along the external cordon. It is the purpose of this chapter to examine in detail the organisation and analysis of home interview surveys.

The home is the major source of trip generation and details of a considerable proportion of trips generated in an urban area may be obtained by this form of survey. A home interview survey usually involves leaving a diary or similar document for the respondent to complete at a later stage.

It is not necessary to obtain details of the trips from every household; instead a sampling procedure may be used. One way in which an unbiased sample may be obtained is by sampling from the electoral roll or from the valuation list, both of which are held by local authorities.

The Department of Transport in their Traffic Appraisal Manual[1] advise that the first step in selecting a sample is to define the 'population' of interest, the households in the area being surveyed. Normally all members of the household are of interest in the survey but it might be car drivers, adults, public transport users or some other category. Having decided upon the population of interest an approximation to the population can be obtained from the Electoral Register or the Valuation List.

14

Sample size must depend upon the errors in the data collection process and in the subsequent trip prediction process. Where approximate estimates of travel

Population size	Sample size %
under 50 000	20
50 000–150 000	12.5
150 000–300 000	10
300 000–500 000	6.66
500 000–1 000 000	5
over 1 000 000	

requirements are necessary, a smaller sample size may be tolerated. The Institute of Traffic Engineers[2] in its Manual of Traffic Engineering Studies recommends sample sizes based on the population of the study area.

The Traffic Appraisal Manual[1] gives the following equation for the calculation of sample size, n

$$n = \frac{P(1-P)N^3}{(\frac{E}{1.96})^2 (N-1) + P(1-P)N^2}$$

where N is the total number of households within the survey area,
E is the required accuracy expressed as a number of households,
P is the proportion of households with the attribute of interest.

For example in a household-interview survey it is estimated that there are approximately 35 500 households and it is also estimated that not more than 1 in 20 households will, on the survey date, make the particular journey of which the characteristics are to be modelled. If the required accuracy is 2 per cent of the population calculate the required sample size.

From the above equation

$$n = \frac{0.05 \times 0.95 \times 35\,500^3}{(\frac{710}{1.96})^2 (35\,499) + 0.05 \times 0.95 \times 35\,500^2}$$

$$= 450.$$

Care should be taken to obtain an unbiased sample and if a completed questionnaire cannot be obtained or if it is impossible to deliver a questionnaire then these questionnaires should be omitted from the survey and not delivered at adjacent homes. The trip information obtained must be increased to give all the trips in a zone by the use of a sampling factor

$$\text{sampling factor} = \frac{A - (C \times A/B)}{B - (C + D)}$$

where A = total households in zone
      B = number of households selected for sampling in zone
      C = number of households vacant or demolished
      D = number of households refusing to co-operate

As it is desired to obtain relationships between tripmaking and such variables as land-use characteristics and socio-economic factors, many details other than those relating to the trips must be obtained. The cost of the survey is considerable. Hence great care must be taken in the preparation of these questionnaires both from the point of view of obtaining the required information and also of coding and analysis.

There are alternative approaches to the collection of this data. An interviewer may attempt to contact every member of a selected household individually and details of tripmaking and socio-economic characteristics of the tripmaker are recorded in the presence of the interviewer. Alternatively the interviewer may only collect details of the household characteristics, leaving the questionnaire on trip-making to be completed at a future date. The interviewer will then collect the questionnaire in the future after checking them for completeness. While the first method should result in accurate detailing of tripmaking, it is expensive and there is a danger of interviewer prompting. The second method is less expensive in inter-viewer costs but will tend to be less accurate. Figure 3.1 shows a typical question-naire used to record the characteristics both of the dwelling unit and the residents while details of the trips they make are recorded on the type of form shown in figure 3.2. Both the dwelling unit summary sheet and the trip information sheet are based on reference 3.

Advice on the successful conduct of home-interview surveys is given by the Department of Transport in the Traffic Appraisal Manual. It is stressed that as

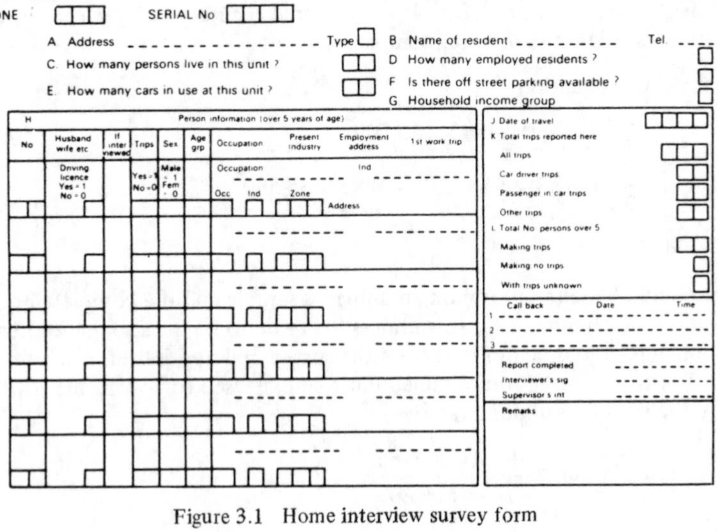

Figure 3.1   Home interview survey form

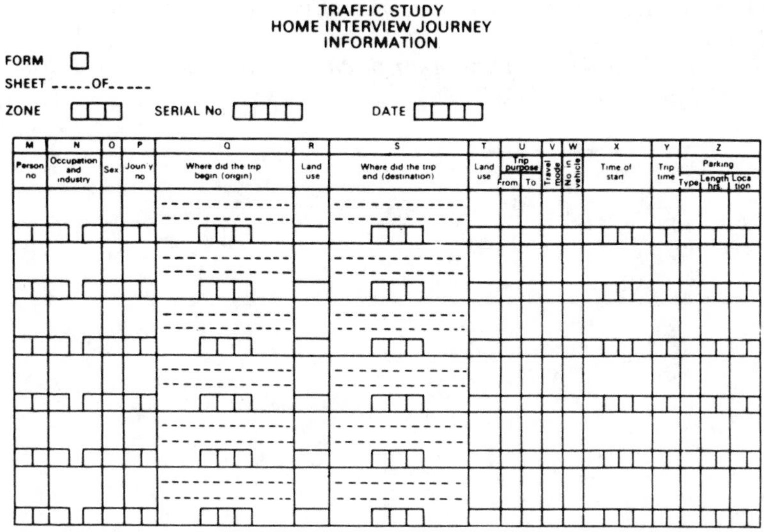
TRAFFIC STUDY
HOME INTERVIEW JOURNEY
INFORMATION

FORM ☐
SHEET -----OF-----
ZONE ☐☐☐ SERIAL No ☐☐☐☐ DATE ☐☐☐☐

M	N	O	P	Q	R	S	T	U		V	W	X	Y	Z		
Person no	Occupation and industry	Sex	Journey no	Where did the trip begin (origin)	Land use	Where did the trip end (destination)	Land use	Trip purpose		Travel mode	No in vehicle	Time of start	Trip time	Parking		
								From	To					Type	Length hrs	Location

Land use codes (R)
1 Offices 5 Public buildings
2 Industry 6 Utilities, etc
3 Shops, etc 7 Open space
4 Residential 8 Vacant land

Purpose codes (U)
1 Home 5 Social/recreation
2 Work 6 Personal business
3 Shop 7 Change mode
4 School 8 Other

Mode codes (V)
1 Car driver 5 Goods vehicle
2 Car pass 6 Motor cycle
3 Bus 7 Other
4 Taxi

Parking codes (Z)
1 Kerbside free 5 Paid park
2 Kerb restricted 6 Not parked
3 Kerb meter 7 Private park
4 Free park 8 Residence

Figure 3.2 Home interview journey information

many home-interview surveys require the householder to recall past details of trip-making it is essential that the interviewer ensures that trip details are accurately remembered by the respondent. A successful interviewer is also required to make sure that the respondent to the travel survey understands his part in the transportation survey. For any survey to be successful it is necessary for the respondents who supply the trip information to have a strong motivation to answer the questions correctly; hence the interviewer should aim to increase the respondent's motivation.

In a transportation study recently carried out for Baghdad by Scott Wilson Kirkpatrick and Partners the home interview survey formed the major element of the travel surveys for the Study Area. Its aim was to collect detailed travel information on the tripmaking characteristics of a 3 per cent sample of the total number of households in the Study Area. Information was obtained on the journeys made on the day before the interview and on the previous Friday by all members of the sample household who were over the age of four. Questions asked during the interview covered household statistics, personal data on residents and visitors staying overnight and trip data. The sample of households was selected from Electricity Board records, 1977 Census household data and from maps where significant development had taken place since 1977. A total of 17 266 interviews were carried out, 15 698 at private households, 1115 at hotels and 453 at university accommodation.

References

1. Department of Transport, *Traffic Appraisal Manual*, London (1982)

2. Institute of Traffic Engineers, *Manual of Traffic Engineering Studies*, Washington DC, (1964)
3. Royal Borough of New Windsor, *A traffic and parking study for the Royal Borough of New Windsor* (1972)

4

The external cordon and screenline surveys

The object of the external cordon survey is to obtain information on the trips originating outside the external cordon having origins within the external cordon or passing through the survey area.

For highway trips whether by public or private transport this information is obtained by an origin–destination survey carried out at census points on the roadside. For non-highway public transport trips a survey is carried out while the trips are being made or, alternatively, an examination of ticket records may give some of the required origin–destination information.

Origin–destination surveys at the external cordon may be carried out by interviewing drivers at census points. Alternatively, in certain limited circumstances, the survey may be accomplished by using questionnaire postcards handed to drivers at the census points for later completion and return by post.

The first method, that of direct interview of drivers, is probably one of the most widely used techniques and is commonly used not only to obtain details of trips which cross the external cordon in transportation studies but also for estimating the journeys which would be made on proposed highways.

Direct interviews at roadside census points are often carried out simultaneously at all the census points but a reduction in the number of staff required at any one time, but not in total, and also in the shelters, barriers and warning signs necessary is possible by dealing with different census points and directions of traffic movement at separate times.

Questions asked at the census points will, of course, vary according to the survey, but they should be carefully phrased and without ambiguity. To avoid bias in the results obtained, the questions may sometimes be printed on a card and shown to a driver but it is more usual for the interviewers to be given instructions to ask the exact question required by the traffic engineer in charge of the survey.

A form is usually completed for each interview, the enumerator noting the time, the census point and the direction being sampled, the class of vehicles, the origin and destination of the journey and any intermediate stops. If the economic benefits of the proposed highway are to be assessed then additional questions may be asked to determine the purpose of the journey and the number of occupants of the vehicles. A typical interview sheet used for recording details of vehicle trips is given in figure 4.1.

Under normal traffic conditions it is of course not possible to interview the driver of each vehicle without causing undue delay or employing an excessive number of interviewers. For this reason only a proportion of the vehicles travelling past a census point are stopped for interview, a process known as sampling.

The selection of the vehicles for the sample must be free from bias and the sampling technique developed by the Transport and Road Research Laboratory has been widely used[1].

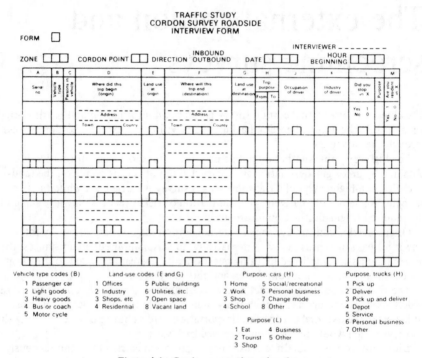

Figure 4.1   Cordon survey interview form

In this method a police officer, upon receiving a signal from either of the two interviewers that he is ready to commence the next interview, directs the first vehicle that arrives at the census point into the interview lane. All other traffic is allowed to pass the census point without being halted, a volume count of the number of vehicles passing the point being carried out by the enumerator.

If the first vehicle to arrive at the census point after an interviewer has indicated that he is ready to carry out a further interview is halted, then the sampling technique is free from bias. It has been found, however, that on derestricted lengths of highway the first vehicle may be travelling at a high speed and the police officer directing the traffic may halt a slower vehicle which is following. Should this tendency not be eliminated by slowing down the whole of the traffic stream then the sample is biased in favour of slower vehicles.

As no attempt is made to ensure that the number of vehicles halted is a constant proportion of the total number of vehicles passing the census point, a sampling factor is introduced for each class of vehicle which varies according to the hour. The sampling factor for a certain hour is simply the number of vehicles of a particular class passing the census point during that hour divided by the number of vehicles of the particular class included in the sample.

The growth of traffic that has occurred since the publication of reference 1 has led to the issue of detailed instructions for the conduct of traffic surveys by roadside interview[2].

General instructions are given on the location of census points with particular emphasis on minimising accident risk, carriageway obstruction and delay to vehicles. It is obviously important to site the census point away from junctions, on lengths of highway without excessive gradient and where the carriageway is as wide as possible. For single carriageway roads it is desirable to incorporate a bypass lane otherwise it will be necessary to operate stop/go or shuttle working. There are fewer difficulties when a census point is located on a dual carriageway road provided speeds and flows are not high. In the British Isles interviewing on motorways has not been allowed and only very occasionally has it been allowed on slip roads.

The information requested at roadside interview surveys carried out in accordance with these recommendations is standardised. The data collected from the drivers sampled comprises: the full postal address of the next stop of the vehicles; the reason for the journey to the next destination which is classified into the categories of home, holiday home, work, employer's business, education, shopping, personal business, visiting friends and recreation/leisure. Details are also required of the last stop of the vehicle and the reason for being at that address; the same categories as are used for destinations are employed. Car, motorcycle and van drivers are asked the full postal address at which the vehicle was garaged during the previous night.

In addition to interviewers it is also necessary for enumerators to undertake classified counts of vehicles passing through the census station during half-hour periods.

Registration number surveys

If it is desired to know the points at which vehicles enter and leave a survey area but it is not desirable to stop vehicles then registration number surveys are carried out. In these surveys the class of each vehicle and its registration number are noted as it passes each survey point together with the time the vehicle was observed. Subsequent comparison of the records will indicate entry and exit points into and from the survey area together with travel times for different classes of vehicles.

Records can be obtained by manual recording and the Traffic Appraisal Manual recommends that one observer be stationed at each station and note as many registration numbers as possible. If because of the traffic flow or for any other reason a number is missed, the observer places a tick on the recording sheet. In this way a record of the traffic flow is obtained. To facilitate recording it is recommended that only a portion of the registration number be noted (that is, for A405NJX record

A405, for DNW405Y record 405Y). Vehicle arrival and departure times should be noted to the nearest minute and a continuous check be kept on the time. Classification of vehicles can be made by the use of an additional code letter.

It is also recommended that control cars be used to control the quality of the information; they pass through the survey area noting any unusual conditions and the exact time when they pass the survey stations. In this way the differing observers are connected together both by time and by the control cars.

A problem associated with partial registration number surveys is the possibility that matches may occur between different vehicles, causing spurious matches. This is particularly a problem when flows are heavy but few vehicles actually travel between entry and exit points of the survey area. An approximate formula for the number of spurious matches is given in the Traffic Appraisal Manual[3].

When traffic flows are heavy, manual recording becomes impracticable and use should be made of portable tape recorders. They are left permanently on and registration numbers together with vehicle class recorded by an observer. If a registration number is missed the passage of a vehicle should be marked by the use of some word such as 'blank' so that a count of vehicles passing the point can be made. Frequent time references should be incorporated in the tape.

Using postcards

The use of prepaid postcards which are handed to drivers as they pass the census points is an inexpensive method of obtaining journey information. Delay to vehicles is small and a considerable amount of information may be obtained from the questions on the postcard[2]. A great deal of advance publicity however is required if a reasonable proportion of postcards are to be returned, and often the number of cards returned has been disappointing. There is also the danger that certain types of vehicle drivers are more likely to complete and return the postcards than others, so giving a bias to the results obtained.

Details of the trips crossing the screenlines within the external cordon are necessary to allow a check to be made on the information obtained by the home-interview, external cordon and goods movement surveys. The total existing trip information is assigned to the present day or base year transport networks and if a discrepancy is found then the survey information is adjusted. Screenline checks are also useful to compare sector-to-sector movements as a check on the trip generation, attraction and distribution models developed from the present day pattern of movement.

References

1. Transport Road Research Laboratory, *Research on Road Traffic*, Chapter 4, HMSO, London (1965)
2. Department of Transport, *Traffic Surveys by Roadside Interview*, Advice Note TA/11/81, London (1981)
3. Department of Transport, *Traffic Appraisal Manual*, London (1982)

5

Other surveys

Commercial vehicle surveys

This survey is designed to measure the trips made by commercial vehicles within the internal area. The data is obtained by issuing drivers with forms on which they record each trip, together with trip origin, destination, purpose and parking details. The size of the sample depends on the size of the survey area and also on the variability of goods-use activity, ranging from 100 per cent for a small town to about 25 per cent for a large area. The population may be obtained from the land-use survey,

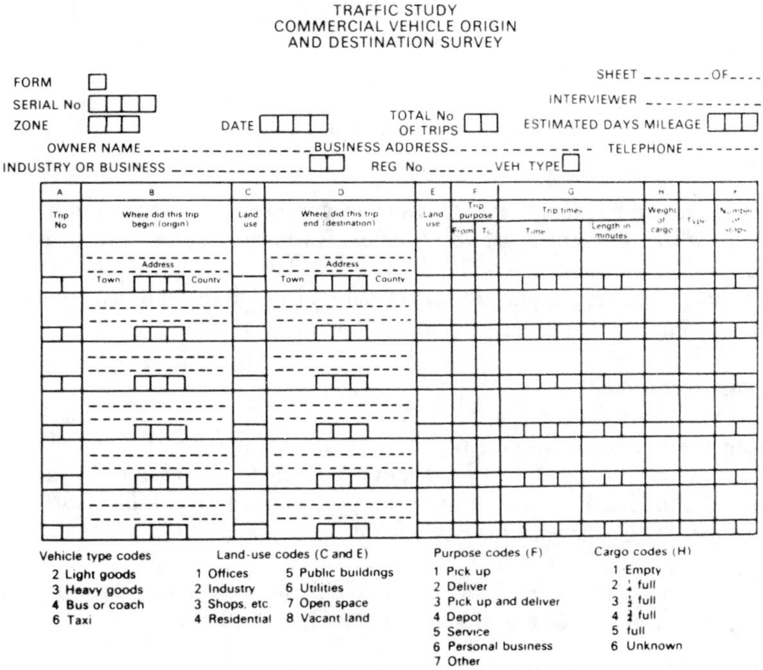

Figure 5.1 Commercial vehicle survey form

23

which will indicate addresses at which commercial vehicles may be kept or alternatively excise licence records may be consulted. A typical questionnaire is reproduced
in figure 5.1.

Bus passenger surveys

Bus trips made by residents are obtained from the home interview survey; trips
made by public transport by non-residents are obtained by a bus passenger survey
carried out at the external cordon. Duplication of trips made by residents within the
survey area should be checked.

Questionnaires are usually distributed to bus passengers as the bus enters the
survey area at the external cordon. The completed questionnaires are collected from
the bus or passengers are requested to return the completed forms by post. The
questionnaire used in the Leicester Traffic Survey[1] is given in figure 5.2 as an
illustration.

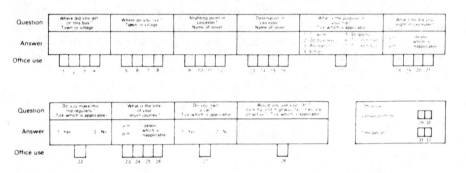

Figure 5.2   A typical bus passenger questionnaire

In the larger urban areas there will also be a considerable number of non-home-
based trips within the central area by tripmakers who are not resident within the
external cordon. As these trips will not be recorded by the home interview survey
it is necessary to either conduct additional surveys within the external cordon or
interview a sample of tripmakers leaving the survey area.

Train passenger surveys

Questionnaires are frequently distributed to train passengers as they board trains
bound for the survey area. Questionnaires are collected on the train or passengers
are requested to return them by post. When postal return is used the information
obtained should be checked wherever possible when the return is less than 70 per
cent.

The information requested on a train passenger survey will vary according to the
survey type. It will include questions requesting information on:

(a) the address of the traveller;
(b) the mode of travel from the home to the station;

(c) the scheduled departure time of the train;
(d) the mode of travel from the destination to the workplace;
(e) the address of the destination;
(f) the journey purpose;
(g) details of vehicle ownership.

Taxi surveys

If required a survey is carried out as for a commercial vehicle survey. Normally taxi trips form only a small proportion of tripmaking in most cities and a taxi survey is not normally necessary unless the number of trips by this mode of transport are considerable.

Reference

1. J. M. Harwood and V. Miller, *Urban Traffic Planning*, Printerhall, London (1964)

Problems

Select the correct answer to the following questions.

1. The commercial vehicle survey obtains:

(a) details of all trips made by commercial vehicles based within the survey area;
(b) details of commercial vehicle trips which cross the external cordon;
(c) an inventory of all commercial vehicles garaged within the external cordon.

2. Bus and train passenger surveys obtain:

(a) details of trips made by residents who live within the external cordon;
(b) details of trips made into and through the survey area by tripmakers resident outside the external cordon;
(c) details of journey times by public transport within the survey area.

Solutions

The correct answers to the above questions are:

1. The commercial vehicle survey obtains:

(a) details of all trips made by commercial vehicles based within the survey area.

2. Bus and train passenger surveys obtain:

(b) details of trips made into and through the survey area by tripmakers resident outside the external cordon.

6
Trip generation

Once the transportation survey has collected all the details of the existing tripmaking pattern and the socio-economic, land-use and trasportation-system characteristics of the survey area, the second stage in the transportation planning process is the development of relationships between the total number of trip origins and destinations in a zone and the zonal characteristics. It is assumed that these relationships will be true in the future and so, if land-use and socio-economic factors can be predicted, future trips can be estimated for any proposed transport system.

At this stage some basic definitions have to be made. A trip is a one-way person movement by one or more modes of travel and each trip will have an origin and a destination. Most surveys divide trips into home-based and non-home-based trips.

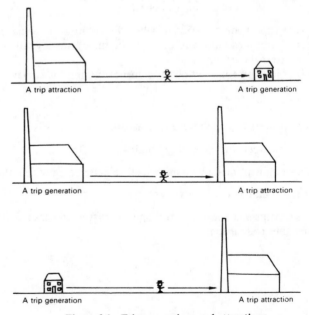

Figure 6.1   Trip generations and attractions

All trips which have one end at the home are said to be generated by the home and
the other end of the trip is said to be attracted to the zone in which it commences
or terminates. For non-home-based trips the zone of origin is said to generate the
trip and the zone of destination is said to attract the trip. Figure 6.1 illustrates the
differences between generation and attraction.

The form of the relationship connecting trip generation and land use and socio-
economic factors depends upon the basic framework of the transportation process.
If the study is designed to use trip end modal-split models, which allocate trips to
differing modes of transport before the distribution of trips between the traffic
zones, then trip generation models are designed to predict person movement by
differing modes of travel. On the other hand when trips are distributed between the
zones before the decision is made regarding the mode of travel, then a trip interchange model is required, which predicts person movement in terms of total person
movement by all modes of travel.

Because trips made for different purposes have different distribution and modal
split characteristics, trip generation equations are generally stratified into trips for
different purposes whether trip end or trip interchange models are being developed.

As the home is regarded as being the generator of the major proportion of trips,
a considerable number of trips have the home end of the trip as part of the description; that is

> home to work trips
> home to recreation trips

and where the land use at the destination does not fall into any of the stated categories then the description of

> home to other trips

is used

Trip generation equations have as their dependent variable the number of trips
generated per person or per household for different trip purposes, while the independent variables are the'land-use and socio-economic factors that are considered to
affect tripmaking.

Land use is of course a major consideration in the generation of trips for in
residential areas the density of development and the type of housing can be expected to be major factors in the number of trips likely to be generated. Similarly
in areas where commercial or industrial land use is of importance then the type of
activity and the number of workers employed per unit area will influence trip
generation.

Other factors connected with the home, that are considered to influence the trip
generation rate, are family size, income, motor vehicle ownership and the social
status of the head of the household being considered.

These relationships between trip ends, land-use and socio-economic factors may
be obtained in two ways:

(1) by zonal least squares regression analysis;
(2) by category analysis.

Zonal least squares regression analysis

Past transportation studies have made extensive use of zonal least squares regression. For each of a number of zones a certain number of trip ends—the dependent variable —are observed and each zone has certain measurable characteristics to which this trip generation rate may be related. These characteristics X_1, X_2 etc. are referred to as the independent variables and are the land-use and socio-economic factors which have been previously referred to.

The equation obtained by least squares analysis is of the general form

$$Y = b_0 + b_1 \times X_1 + b_2 \times X_2 + \ldots + b_n \times X_n$$

where b_0 is the intercept term or constant.

$b_0, b_1 \ldots b_n$ are obtained by regression analysis,

$X_1, X_2 \ldots X_n$ are the independent variables.

In developing regression equations it is assumed that:

1. All the independent variables are independent of each other.
2. All the independent variables are normally distributed; if the variable has a skew distribution often a log transformation is used.
3. The independent variables are continuous.

It is not usually possible for the transportation planner to conform to these requirements and regression analysis has been subjected to considerable criticism.

Nevertheless regression analysis is a powerful tool when used in conjunction with a computer for handling considerable volumes of data.

It is important however to recognise that the regression process contains the likelihood of the future values of the dependent variable Y being in error when future values of the independent variables X_1, X_2, etc., are substituted into the equation.

The likely sources of error may be stated to be:

(a) errors in the determination of the existing values of the independent variables owing to inaccuracy or bias in the transportation survey;
(b) errors in the determination of the existing values of the dependent variables, also as a result of inaccuracy or bias in the transportation survey. This may be detected and corrected by adequate screenline checks.
(c) the assumption that the regression of the dependent variable on the independent variables is linear, a matter of some importance when future values of the independent variables are outside the range of observed values;
(d) errors in the regression obtained owing to the scatter of the individual values and the inadequacy of the data;
(e) difficulties in the prediction of future values of the independent variables, for the future value of the dependent variable will only be as good as the future estimates of the independent variables;
(f) future values of the independent variable will be scattered as are the present values;

(g) the true regression equation may vary with time because factors that exert an influence on tripmaking in the future are not included in the present-day regression equation.

Most computer programs introduce or delete the dependent variables in a stepwise manner. Only variables that have a significant effect on the prediction of the dependent variables are included in the regression analysis.

It is usual to compute the following statistical values to test the goodness of fit of the regression equation.

1. Simple correlation coefficient r which is computed for two variables and measures the association between them. As r varies from -1 to $+1$ it indicates the correlation between the variables. A value approaching ±1 indicates good correlation.

2. Multiple correlation coefficient R which measures the goodness of fit between the regression estimates and the observed data. $100R^2$ gives the percentage of variation explained by the regression.

Transportation studies have produced a considerable number of regression equations and their variety is often confusing. This is partly owing to variations in the form of the independent variables which have been used and also to variations in tripmaking.

A typical equation obtained in the Leicester transportation study[1] was

$$Y_8 = 0.0649X_1 - 0.0034X_3 + 0.0066X_4 + 0.9489Y_1$$

where Y_8 = total trips per household where the head of the household is a junior
          non-manual worker/24 h

    X_1 = family size
    X_3 = residential density
    X_4 = total family income
    Y_1 = cars/household

If all the factors influencing the pattern of movement are correctly identified then it might be expected that trip generation equations obtained in one survey would be applicable to other surveys. Attempts to establish similarity between relationships observed in different surveys have not however met with any great success because of variations in both the dependent and independent variables chosen.

Category analysis

Difficulties with the use of regression equations for the study of trip generation have led to considerable support being given to the use of disaggregate models, that is, models based on the household or the person. Making use of the unexpanded sample data, these models make no reference to zone boundaries and allow considerable flexibility in the selection of alternative zone systems when future trip ends are being predicted.

This approach has become known as category analysis and has been largely developed by Wootton, Pick and Gill[2,3] and has been applied to a considerable number of transport studies. Category analysis uses the household as the fundamental unit of the trip generation process and assumes that the journeys it generates depend on household characteristics and location relative to workplace, shopping and other facilities. Trip generation is measured as the average number of one-way trips generated by a household on an average weekday.

Household characteristics that are readily measured and appear to account for variation in generation both at the present time and in the future are disposable income, car ownership, family structure and size. Location characteristics have proved more difficult to isolate and the characteristic that has found the greatest application is public transport accessibility.

Wootton and Pick[2] classified households according to:

1. Cars owned —(1) None
              (2) 1
              (3) More than 1
2. Income   —Originally 6 classes were proposed ranging from less than £500 to £2500 or more. These classes were modified in the West Midland Study to give an income class up to £10 000. There is usually a difficulty in obtaining sufficient observations in higher income groups and yet it is in these ranges that future prediction is important.
3. Family structure

	Adults employed	Adults not employed
(1)	None	1
(2)	None	More than 1
(3)	1	0, 1
(4)	1	More than 1
(5)	More than 1	0, 1
(6)	More than 1	More than 1

These categories produced 108 household classes and associated with each class is a trip rate. It was proposed that trips be classified by mode of travel and trip purpose.

These are:

Modes   1. Drivers of cars or motor cycles
          2. Public transport passengers
          3. Other passengers (mostly car passengers)
Purpose 1. Work   2. Business 3. Education
          4. Shopping 5. Social   6. Non-home based

where there are thus 6 × 3 = 18 mode and purpose trip combinations.

The basic assumption is that trip rates are stable over time and that the future behaviour of a household can be described by the category into which it falls. The number of households in each category can be estimated by the fitting of mathematical distributions to the observed values of income, car ownership and family structure.

A great advantage of the category analysis technique is that it is possible to estimate household categories from the Registrar General's Census Data using known relationships. Trip generation rates obtained from other surveys are then used subject to a small survey check on the accuracy of the rates. When the cost of large scale home interview surveys is considered, the advantage of this technique can be appreciated.

The computational techniques are also simpler than those required for zonal least squares regression. The use of disaggregate data may also be expected to simulate human behaviour more realistically than zonal values.

A disadvantage of the category analysis technique however is that it is assumed that income and car ownership will increase in the future. The categories with high incomes and car ownership are however the ones which are least represented in the base year data. Moreover they are the categories which are most likely to be used for future estimates of trip generation.

Income distribution

Pick and Gill have shown that income distribution may be represented by a continuous probability density function $Q(x)$ such that the number of households having income x, $a < x < b$, is given by

$$N \int_a^b Q(x) \, dx$$

where N is the number of households in the zone.

The following distribution has been used for $Q(x)$

$$Q(x) = \frac{\alpha^{n+1}}{n!} x^n e^{-\alpha x}$$

where $\alpha = \bar{x}/s^2$
$n = (\bar{x}^2/s^2) - 1$,
x is the mean income,
s^2 is the standard deviation of income.

Future incomes were projected on the basis of

$$x^1 = x(1 + g)^y$$

where g is the annual growth rate of income relative to the cost of living,
y is the number of years projected,
x^1 is the future income,
x is the present income.

Car ownership distribution

A conditional probability function $P(n/x)$ was derived from the West Midlands Transportation Study to give the probability of a household owning n cars if its income relative to the price of cars is x.

Then

$$P(0/x) = Ke^{-\beta x}$$

$$P(1/x) = Ce^{-\beta x} \times (\beta x)^n$$

where K and C are constants and β varies with bus accessibility.

Variations of $P(0/x)$ and $P(1/x)$ within a study area as given by Pick and Gill are shown in figure 6.2. These variations have been explained by the following factors, but more detailed information can be obtained from reference 3.

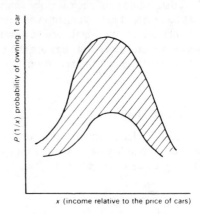

Figure 6.2   Probability of car ownership (adapted from ref. 3)

(a) An increase in residential density produces a higher $P(0/x)$ curve and a lower $P(1/x)$ curve. It has been found that the effects of high and low density are well defined but there is some variation in the medium density ranges

(b) An increase of the public transport accessibility index of a traffic district causes an increase in the value of $P(0/x)$. It is defined as

$$\sum_{j} \sqrt{b_j} / \sqrt{a_i}$$

where a_i = area of district i (sq. miles)

b = number of buses on route j passing through the district per unit time.

This index has a value of 0 in a rural area without bus services and a value of approximately 60 in central London.

(c) The cost of housing has been bound to influence car ownership: the greater the housing cost the higher the $P(0/x)$ curve.

(d) Lack of garage space inhibits multi-car owners and influences both the $P(0/x)$ and $P(1/x)$ curve.

(e) Because older household members are likely to have more money available for car purchase and younger household members are likely to be moped and motor cycle owners, age structure influences car ownership curves.

(f) Spatial relationships affect car ownership for if activities are closely related to housing then there is less necessity for the purchase of vehicles.

More recent work by the Department of Transport for the development of the Regional Highway Traffic Model[4] based on Family Expenditure Survey data for 1965, 1966, 1969-75 indicated that a relationship stable over time existed between household income deflated by an index of car price. Doubts concerning this relationship were raised when Family Expenditure Survey data for 1976-78 became available which indicated car ownership would fall because car prices rose more rapidly than real gross household incomes, but in actual fact car ownership continued to rise. For this reason a new model was proposed which did not include the car price index. Instead the independent variables defining car availability were replaced by the gross household income deflated by the Retail Price Index and driving licences per adult. It is hoped that the latter variable will reflect behaviour over time.

Family structure distribution

As previously stated households are split into 6 categories and the number of households in each category may be estimated using the following argument.

Let the probability of a household having n members be $Q(n)$.

Let $p(n)$ be the probability that a member of an n member household is employed. The probability that r members of an n member household are employed is

$$\frac{n!}{r!(n-r)!} p(n)^r (1 - p(n))^{n-r}$$

assuming a binomial distribution. The probability that a household has n members of whom r are employed is then

$$P(n, r) = \frac{n!\, Q(n)}{r!\, (n-r)!}\, p(n)^r (1 - p(n))^{n-r}$$

where $Q(n) = \dfrac{e^{-x} x^{n-1}}{(n-1)!}$ $n = 1, 2, \ldots$

x = average family size $-$ 1

$p(n) = \dfrac{\text{employed residents}}{\text{total households}}$

Trip generation and attraction models developed for the Regional Highway Traffic Model

The Department of Transport[4] has developed models for two basic types of trip: home-based trips which have either origin or destination at the home, and non-home-based trips which have neither origin nor destination at the home.

Home-based trip end models were developed for three separate trip purposes: home-based work trips, home-based employer's business and home-based other. For non-home-based trips models were developed for employer's business and other trips.

For both home-based and non-home-based trip models two separate but related sub-models were calibrated. In the case of home-based purposes the models were for trip generation and trip attraction, for non-home-based trips models were developed for trip generation (or creation if the purpose was non-home-based) and trip attraction (or allocation if the purpose was non-home-based).

Both models for non-home-based trip creations had as the independent variables the proportions of households in the zone with zero, one, and two or more cars. The models for non-home-based trip allocations had as independent variables the households in the zone, total employment places retail employment places, manufacturing employment places and service employment places in the zone and the average car ownership per household.

The generation models developed were:

$G_w = (a + bE_y)(1 - P)$ work car trip generations per household for car ownership group Y and zone type Z.

$G_{eb} = (a + bE_y)P$ employer's business car trip generations per household for car ownership group Y and zone type Z.

$G_0 = c + dL + eN$ other purpose car trip generations per household for car ownership group Y and zone type Z.

where E_y is the average number of employed residents per household in car ownership group Y and zone type Z.

P is the proportion of work and employer's business trips which are employer's business in car ownership group Y and zone type Z.

L is the average number of driving licences per household in car ownership group Y and zone type Z.

N is the average number of non-employed residents per household.

and a, b, c, d and e are coefficients which are calibrated for car ownership group Y and zone type Z.

The average number of employed residents in one car and multi-car owning households (E_1 and E_2) was determined by:

$E_1 = E + B(1 - C)$

$E_2 = E + B(2.169 - C)$

where E and C were respectively the zonal averages of Employed Residents and Cars per household and the form of this equation with $B = 0.62$ ensured compatibility with the overall planning data estimates of activity rates. The value of 2.169 was the average number of cars per household in multi-car owning households.

Eight different types of zones (Z) were defined, comprising 2 types of rural zones, 3 types of urban zones and 3 types of London zones. Three levels of household car ownership (Y) were also defined, zero, one and two or more cars available per household.

The home-based attraction models which were developed were for work trips

$$A_w = f \times TE + g \times TE \times AX + h \times AREA$$

for employers business trips

$$A_{eb} = i \times TE \times AX$$

for other trips

$$A_0' = j \times RET \times AX + k \times HH \times AX + l \times CARS$$

where TE is the total employment in the zone

   AX is a measure of car accessibility
   $AREA$ is the zone area in hectares
   RET is the retail employment in the zone
   HH are the households in the zone
   $CARS$ is the number of cars in the zone
   f, g, h, i, j, k, l are calibration coefficients

The category analysis method of modelling trip generation was employed by Scott Wilson Kirkpatrick and Partners for home-based trip productions in the Baghdad Transportation Study.

The household categories used were based on five variables; these were, number of persons per household, number of workers per household, number of cars available per household (0, 1 or 2 plus), income per household (low, medium or high) and household density (less than 1, and more than 20 and 1 to 20 households per hectare). The variables of number of workers per household and number of persons per household were combined into 10 worker/resident groups. The variable of household density was added to the analysis when it was found that low-density rural areas and also high-density urban areas had trip rates which were lower than the average.

Household trip rates were found to increase rapidly with increasing household income and with increasing car availability. On the other hand, the trip rate did not vary greatly with household size, especially in the range 5 to 11 residents per household.

Future numbers of households in each category were predicted from a household category generation sub-model and the future trip generation from these future households obtained by assuming present day trip generation rates remained constant with time.

For non-home-based trips, stepwise multiple linear regression analysis was used to develop trip generation equations.

An example of a trip generation equation derived by stepwise linear regression analysis was for non-home-based zonal attractions which had the form:

24 hour NHB attractions = 130.1
$$+ 4.209 \text{ Employment in wholesale trade}$$
$$+ 0.907 \text{ Employment in community services}$$
$$+ 2.337 \text{ Number of restaurant seats}$$
$$+ 0.765 \text{ Employment in retail trade}$$
$$+ 0.729 \text{ Employment in construction}$$

References

1. J. M. Harwood and V. Miller, *Urban Traffic Planning*, Printerhall, London (1964)
2. H. J. Wootton and G. W. Pick, Travel estimates from census data, *Traff. Engng Control*, 9 (1967), 142–5
3. G. W. Pick and J. Gill, New developments in category analysis, *PRTC Symposium, London* (1970)
4. Department of Transport, *Traffic Appraisal Manual*, London (1982)

Problems

Are the following statements true or false?

(a) A trip with an origin at the workplace and a destination at the home is said to be generated by the home.

(b) A journey from work to home made by walking to the bus, travelling by bus to the station and completing the journey by train is regarded as three trips.

(c) A trip end modal-split generation model predicts the trips generated by a traffic zone regardless of the mode of travel.

(d) In a trip generation equation the independent variables usually describe land-use and socio-economic factors.

(e) There would be no objection to the use of total household income and number of employed members in the household as independent variables in a trip generation equation.

(f) The use of trip generation equations to predict future trips depends on the ability to estimate future values of the independent variables.

(g) Category analysis uses disaggregate survey data while regression analysis employs zonal aggregate survey data.

(h) The examination of survey data by the use of category analysis techniques shows that the trip generation rate decreases with car ownership and increases with income class.

Solutions

Statement (a) is correct since all work trips with an origin or a destination at the home are said to be generated at the home.

Statement (b) is incorrect since a trip is a single journey between an origin and a destination.

Statement (c) is incorrect since a trip end modal-split model predicts trips classified by modal type; that is, modal split is carried out before trip distribution takes place.

Statement (d) is correct because trip generation equations have as their dependent variable the number of trips generated while the independent variables are the land-use and socio-economic factors which affect the generation of trips.

Statement (e) is incorrect in that the independent variables should be independent of each other while there is likely to be a strong correlation between the number of employed members of a household and the total household income.

Statement (f) is correct because future trips are estimated by assuming that the same correlation as exists today between tripmaking and land-use and socio-economic factors will exist in the future. Future trips are then estimated by substituting future land-use and other factors into the trip generation equations.

Statement (g) is correct in that category analysis attempts to predict tripmaking at the household level while least squares regression analysis derives equations from aggregated zonal data.

Statement (h) is incorrect since the analysis of the tripmaking habits of households shows that the number of trips made by households increases as the number of cars owned by the household increases and as household income increases.

7

Trip distribution

Trip distribution is another of the major aspects of the transportation simulation process and although generation, distribution and assignment are often discussed separately, it is important to realise that if human behaviour is to be effectively simulated then these three processes must be conceived as an interrelated whole.

In trip distribution, two known sets of trip ends are connected together, without specifying the actual route and sometimes without reference to travel mode, to form a trip matrix between known origins and destinations.

There are two basic methods by which this may be achieved:

1. Growth factor methods, which may be subdivided into the

   (a) constant factor method;
   (b) average factor method;
   (c) Fratar method;
   (d) Furness method.

2. Synthetic methods using gravity type models or opportunity models.

Trip distribution using growth factors

Growth factor methods assume that in the future the tripmaking pattern will remain substantially the same as today but that the volume of trips will increase according to the growth of the generating and attracting zones. These methods are simpler than synthetic methods and for small towns where considerable changes in land-use and external factors are not expected, they have often been considered adequate.

(a) *The Constant Factor Method* assumes that all zones will increase in a uniform manner and that the existing traffic pattern will be the same for the future when growth is taken into account. This was the earliest method to be used, the basic assumption being that the growth which is expected to take place in the survey area will have an equal effect on all the trips in the area. The relationship between present and future trips can be expressed by

$$t'_{ij} = t_{ij} \times E$$

where t'_{ij} is the future number of trips between zone i and zone j. t_{ij} is the present number of trips between zone i and zone j. E is the constant factor derived by dividing the future number of trip ends expected in the survey area by the existing number of trip ends.

This method suffers from the disadvantages that it will tend to overestimate the trips between densely developed zones, which probably have little development potential, and underestimate the future trips between underdeveloped zones, which are likely to be extremely developed in the future. It will also fail to make provision for zones which are at present undeveloped and which may generate a considerable number of trips in the future.

(b) *The Average Factor Method* attempts to take into account the varying rates of growth of tripmaking which can be expected in the differing zones of a survey area.

The average growth factor used is that which refers to the origin end and the destination end of the trip and is obtained for each zone as in the constant factor method. Expressed mathematically, this can be stated to be

$$t'_{ij} = t_{ij} \frac{(E_i + E_j)}{2}$$

where $E_i = \dfrac{P_i}{p_i}$ and $E_j = \dfrac{A_j}{a_j}$

t'_{ij} = future flow ab,

t_{ij} = present flow ab,

P_i = future production of zone i,

p_i = present production of zone i,

A_j = future attraction of zone j

a_j = present attraction of zone j

At the completion of the process attractions and productions will not agree with the future estimates and the procedure must be iterated using as new values for E_i and E_j the factors P_i/p'_i and A_j/a'_j where p'_i and a'_j are the total productions and attractions of zones i and j respectively, obtained from the first distribution of trips. The process is iterated using successive values of p'_i and a'_j until the growth factor approaches unity and the successive values of t'_{ij} and t_{ij} are within 1 to 5 per cent depending upon the accuracy required in the trip distribution.

The average factor method suffers from many of the disadvantages of the constant factor method, and in addition if a large number of iterations are required then the accuracy of the resulting trip matrix may be questioned.

(c) *The Frator Method*[1] This method was introduced by T. J. Fratar to overcome some of the disadvantages of the constant factor and average factor methods. The Fratar method makes the assumptions that the existing trips t_{ij} will increase in proportion to E_i and also in proportion to E_j. The multiplication of the existing flow by two growth factors will result in the future trips originating in zone i being greater than the future forecasts and so a normalising expression is introduced which is the sum of all the existing trips out of zone i divided by the sum of all the existing trips out of zone i multiplied by the growth factor at the destination end of the trip. This may be expressed as

$$t'_{ij} = t_{ij} \times \frac{P_i}{p_i} \times \frac{A_j}{a_j} \times \frac{\Sigma^k t_{ik}}{\Sigma^k_{(A_k/a_k)} t_{ik}}$$

where t'_{ij} = future traffic flow from i to j

t_{ij} = existing traffic flow from i to j

P_1 and A_j are total future trips produced by zone i and attracted to zone j respectively.

p_i and a_j are total existing trips produced by zone i and attracted to zone j respectively.

k = total zones.

The procedure must be iterated by substituting t'_{ij} for t_{ij}, $\sum_j t'_{ij}$ for p_i, $\sum_i t'_{ij}$ for a_j.

Agreement to between 1 and 5 per cent is achieved by successive iterations.

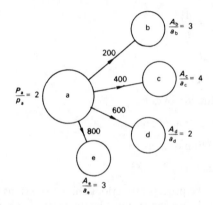

Figure 7.1 The Fratar method of trip distribution

The Fratar method can be illustrated by the simple example shown in figure 7.1 which shows the growth factors and the existing trip pattern.

Then

$$t'_{ij} = t_{ij} \times \frac{P_i}{p_i} \times \frac{A_j}{a_j} \times \frac{\Sigma^k t_{ik}}{\Sigma^k \left(\frac{A_k}{a_k}\right) t_{ik}}$$

that is

$$t'_{ab} = 200 \times 2 \times 3 \times \frac{(200 + 400 + 600 + 800)}{200 \times 3 + 400 \times 4 + 600 \times 2 + 800 \times 3}$$

$$= 414$$

$$t'_{ac} = 400 \times 2 \times 4 \times \frac{(200 + 400 + 600 + 800)}{200 \times 3 + 400 \times 4 + 600 \times 2 + 800 \times 3}$$

$$= 1103$$

$$t'_{ad} = 600 \times 2 \times 2 \times \frac{(200 + 400 + 600 + 800)}{200 \times 3 + 400 \times 4 + 600 \times 2 + 800 \times 3}$$

$$= 828$$

$$t'_{ac} = 800 \times 2 \times 3 \times \frac{(200 + 400 + 600 + 800)}{200 \times 3 + 400 \times 4 + 600 \times 2 + 800 \times 3}$$

$$= 1655$$

In this example the future trips produced by zone a meet the requirements that $P_a/p_a = 2$, but the requirements that

$$\frac{A_b}{a_b} = 3; \quad \frac{A_c}{a_c} = 4; \quad \frac{A_d}{a_d} = 2 \quad \text{and} \quad \frac{A_c}{a_c} = 3$$

are not met and in a practical example, where the number of zones would be considerably greater, further iterations would be required.

(d) *The Furness Method*[2]. In this method the productions of flows from a zone are first balanced and then the attractions to a zone are balanced. This may be expressed

$$t'_{ij} = t_{ij} \times \frac{P_i}{p_i}$$

$$t''_{ij} = t'_{ij} \times \frac{A_j}{\Sigma \text{ trips attracted to } j \text{ in first iteration}}$$

$$t'''_{ij} = t''_{ij} \times \frac{P_{ij}}{\Sigma \text{ trips produced by } i \text{ in second iteration}}$$

where the symbols are as previously stated.

Usually these simplified approaches to trip distribution are only suitable for smaller surveys where similar growth factors are applied to zones and where considerable areas of new development are not expected. The following extract from *Land Use/ Transport Studies for Smaller Towns*[3] illustrates the use of growth factors and the Furness method of trip.

The four steps in the calculation are:

1. Total the outgoing trips for each zone and multiply by the zonal growth factor to obtain the predicted origin outgoing totals.
2. Multiply lines in the matrix by the appropriate origin factor.
3. Total the incoming trips into each zone and divide into the predicted incoming totals to obtain the destination factors.

4. Repeat the iteration processes until the origin or destination factor being calculated is sufficiently close to unity (within 5 per cent is normally satisfactory).

Example

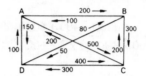

Figure 7.2   The Furness method of trip distribution (initial flows)

TABLE 7.1   Present flows

		Destinations				Present outgoing totals	Predicted outgoing totals	Growth (origin) factor
	Origin	A	B	C	D			
	1	2	3	4	5	6	7	8
9	A	0	200	500	150	850	2550	3.0
10	B	100	0	300	50	450	1125	2.5
11	C	200	200	0	300	700	1400	2.0
12	D	100	80	400	0	580	925	1.6
13	Present incoming totals	400	480	1200	500		6000	
14	Predicted incoming totals	480	720	3600	1200		6000	
15	Acceptance (destination) factor	1.2	1.5	3.0	2.4			

Notes: col. 7 = col. 6 × col. 8
line 14 = line 13 × line 15
Total col. 7 must equal approximately total line 14; that is future trip ends in system must balance, any adjustment to calculated acceptance factors necessary to secure balance being made at the non-residential end of the trips.

TABLE 7.2 Iteration 1

		A	B	C	D	
		16	17	18	19	
20	A	0	600	1500	450	Figures in this matrix
21	B	250	0	750	125	are those in lines 9–12
22	C	400	400	0	600	multiplied by respec-
23	D	160	128	640	0	tive origin factors in
						column 8
24		810	1128	2890	1175	
25		480	720	3600	1200	As in line 14
26	New	0.59	0.64	1.25	1.02	
	destination					
	factors					

TABLE 7.3 Iteration 2

	A	B	C	D		(as Col. 7)	
A	0	384	1870	458	2712	2550	0.94
B	148	0	935	128	1211	1125	0.93
C	217	256	0	612	1105	1400	1.26
D	95	82	797	0	974	925	0.95

Figures in this matrix are columns 16–19 multiplied by respective new destination factors on line 26.

TABLE 7.4 Iteration 3

	A	B	C	D
A	0	362	1760	432
B	138	0	870	119
C	300	324	0	775
D	90	78	752	0
	528	764	3382	1326
	480	720	3600	1200
	0.91	0.94	1.07	0.91

TABLE 7.5 Iteration 4

	A	B	C	D			
A	0	441	1882	392	2615	2550	0.98
B	125	0	906	108	1141	1125	0.99
C	273	305	0	702	1280	1400	1.09
D	82	73	807	0	962	925	0.96

TABLE 7.6   Iteration 5

	A	B	C	D
A	0	333	1840	383
B	123	0	895	107
C	299	334	0	770
D	79	70	777	0
	501	737	3512	1260
	480	720	3600	1200
	0.96	0.96	1.02	0.95

When all the origin or destination factors being calculated in one iteration are within 5 per cent of unity, the result may be considered satisfactory.

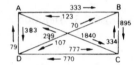

Figure 7.3   The Furness method of trip distribution (distributed flows)

General comment on growth factor methods

The usual application of growth factor methods is in updating recent origin–destination data where the time scale is short and where substantial changes in land use or communications are not expected or have not already taken place. The use of a growth factor method is largely dependent on the accurate calculation of the growth factor itself and this is a likely source of inaccuracy. It is however the lack of any measure of travel impedance which is the major disadvantage of these methods; without this it is impossible to take into account the effect of new and improved travel facilities or the restraining effects of congestion.

Trip distribution using synthetic models

The use of synthetic distribution models allows the effect of differing planning strategies and communication systems and, in particular, travel cost to be estimated, whereas growth factor methods base future predictions on the existing pattern of movement.

    The models are usually referred to as synthetic models because existing data is analysed in order to obtain a relationship between tripmaking and the generation and attraction of trips and travel impedance. These models seek to determine the causes of present day travel patterns and then assume that these underlying causes will remain the same in the future. The most widely used trip distribution model is the so-called 'gravity model'. It has been given this name because of its similarity with the gravitational concept advanced by Newton. It states that trip interchange between zones is directly proportional to the attractiveness of the zones to trips and inversely proportional to some function of the spatial separation of the zones.

The gravity model may generally be stated as

$$t_{ij} = kA_iA_jf(Z_{ij})$$

where A_i and A_j are measures of the attractiveness of the zones of origin and destination to tripmaking and $f(Z_{ij})$ is some function of deterrence to travel expressed in terms of the cost of travel, travel time or travel distance between zones i and j.

If $f(Z_u)$ is taken as $1/Z_{ij}^2$ then the formula is similar to the law of gravitational attraction and for this reason the model is referred to as the gravity model.

Tanner[4] has shown that this form of the deterrence function cannot give valid estimates of travel over large or small distances. He suggested a function of the form

$$\exp(-\lambda Z_{ij}) \times Z_{ij}^{-n}$$

where λ and n are constants.

For work journeys a typical value of λ was found to be 0.2 where Z_{ij} was measured in miles. For many purposes n may be taken as 1.

As Z_{ij} increases the effect of $\exp(-\lambda Z_{ij})$ decreases and the effect of Z_{ij}^n increases.

This is the distribution of trip costs, times or distances which is observed in most studies of tripmaking. The trips that are least in cost, distance or time occur most frequently while trips that are greater in cost, distance or time are observed less frequently. There is considerable evidence that the value of the deterrence function varies with trip type and for this reason practically all studies use differing values for differing trip categories. It is also probable that the function varies with time, but this effect is at present largely ignored.

In practice tripmaking between zones is assumed to be proportional to the total trips generated by the zone of origin of the trip and the total trips attracted by the zone of destination of the trip, or

$$t_{ij} = KP_iA_jf(Z_{ij})$$

where K is a constant.

P_i is the total trip production of zone i, that is $\sum_j t_{ij}$

A_j is the total trip attraction of zone j, that is $\sum_i t_{ij}$

The value of K is frequently taken as

$$\frac{1}{\sum_j A_jf(Z_{ij})}$$

so that

$$t_{ij} = \frac{P_iA_jf(Z_{ij})}{\sum_j A_jf(Z_{ij})}$$

This if often referred to as a production constrained model because the use of this form K results in the total trip production of zone i; that is $\sum_j t_{ij}$ is equal to P_i.

When it is desired to doubly constrain the distribution model, that is

$$\sum_j t_{ij} = P_i$$

$$\sum_i t_{ij} = A_j$$

then K may be replaced by $c_i d_j$, giving

$$t_{ij} = c_i d_j P_i A_j f(Z_{ij})$$

and

$$\sum_j t_{ij} = \sum_j c_i d_j P_i A_j f(Z_{ij}) \quad \text{or} \quad c_i = \frac{1}{\sum_j d_j A_j f(Z_{ij})}$$

and similarly

$$d_j = \frac{1}{\sum_i c_i P_i f(Z_{ij})}$$

The distribution is constrained by setting d_j equal to unity, and obtaining

$$t_{ij} = \frac{d_j P_i A_j f(Z_{ij})}{\sum_j d_j A_j f(Z_{ij})}$$

If $\sum_i t_{ij}$ is not equal to A_j then d_j is made equal to $A_j / \sum_i t_{ij}$ and the procedure iterated until $\sum_i t_{ij}$ approaches A_j.

Past experience has shown that the exponent of travel time varies with travel purpose and also with travel time. For this reason the gravity model is usually presented in the revised form

$$t_{ij} = \frac{P_i A_j F_{ij} K_{ij}}{\sum_{j=1}^n A_j F_{ij} K_{ij}}$$

where F_{ij} = an empirically derived travel time or friction factor which expresses the average area wide effect of spatial separation on trip interchange between zones which are z_{ij} apart. This factor approximates $1/Z^n$ where n varies according to the value of Z expressed as the travel time between zones.

K_{ij} = a specific zone-to-zone adjustment factor to allow for the effect on travel pattern of defined social or economic linkage not otherwise accounted for in the gravity model formulation.

Standard computer programs are available for determining the most suitable values of F_{ij} and K_{ij} for the particular transportation study. These values are then assumed to remain constant in the future to allow future flows to be predicted; this procedure is known as calibration.

The travel time factors are calculated on a survey wide basis by assuming a value of F_{ij} for a given time range. t_{ij} is calculated for all the flows within the same time range and compared with observed values. The procedure is iterated until agreement is obtained and finally zone-to-zone agreement is obtained by the use of K factors.

Application of the gravity model to the London Transportation Study

The gravity model concept was used for trip distribution in the London Transportation Study[12]. It was assumed that the number of trips generated in one district and attracted to another was a function of trip generation and attraction of the two districts respectively and the travel time between them.

These functions, referred to as distribution functions, were calculated from the survey data and assumed to remain constant with time. Trips were stratified by purpose and mode into six types: work, other home-based and non-home-based, for both car owner and public transport trips. Because these functions may vary with trip length the trips were divided into thirteen trip time-length intervals so that approximately equal numbers of trips were in each interval.

Using travel data from the household interviews the distribution functions F_{kt} for each district of attraction k and time interval t were calculated

$$F_{kt} = \frac{\sum\limits_{i(t)} T_{ik}}{A_k \sum\limits_{i(t)} G_i}$$

where i   = district of generation,

T_{ik} = trips generated in district i and attracted to district k,

A_k = total trips attracted to district k,

G_i = total trips generated in district i,

$\sum\limits_{i(t)}$ = summation over all districts of generation, falling in time interval t.

It was found that when these functions were plotted to a logarithmic scale against time, they showed an approximately linear relationship but with a discontinuity in gradient at a trip time length of around 25 minutes.

Forecasting of future trips was carried out separately for each trip purpose after the removal of a fixed percentage of intra-district trips. The first iteration of the distribution was obtained by the use of the model

$$T_{1ik} = \frac{F_{ik} A_k G_i}{\sum\limits_{k} (F_{ik} A_k)}$$

where F_{ik} = distribution function for trips from districts i to k using 1981 trip times,

A_k = forecast attractions in district k

G_i = forecast generations in district i

A ratio was then obtained for each district of attraction, so that

$$R_{1k} = \frac{A_k}{\sum\limits_i T_{1ik}}$$

Using this ratio, a second iteration was made

$$T_{2ik} = \frac{F_{ik}R_{1k}A_kG_i}{\sum\limits_k (F_{ik}A_kR_{1k})}$$

followed by the calculation of

$$R_{2k} = \frac{R_{1k}A_k}{\sum\limits_i T_{2ik}}$$

The procedure was carried out for a third time, the criterion for convergence being that the average (unsigned) error should be within the range 2 to 5 per cent.

Other synthetic methods of trip distribution

While the gravity model has found considerable application in the distribution of generated trips there have been attempts to use other mathematical models which reflect the motivations of tripmakers more closely. Other synthetic trip distribution models that have been used in transportation studies are:

(i)   the intervening opportunities model[5];
(ii) the competing opportunities model[6];

In the intervening opportunities model, which was first used in connection with the Chicago Area Transportation Study[7], the basic assumption is that all trips will remain as short as possible, subject to being able to find a suitable destination. The competing opportunities model has been evaluated in a study of travel patterns observed in the City of Lexington and Fayette County, Kentucky[8]. It differs from the intervening opportunities model in that the adjusted probability of a trip ending in a zone is the product of two independent probabilities, the probability of a trip being attracted to a zone and the probability of a trip finding a destination in that zone.

The intervening opportunities model may be derived as follows

$$t_{ij} = A_i[\exp(-LD) - \exp(-LD_j + 1)]$$

and the probability that a trip will terminate by the time D possible destinations have been considered is

$$P(D) = 1 - \exp(-LD)$$

Then t_{ij} is equal to P_i multiplied by the probability of a trip terminating in j.
   In this treatment L is a constant representing the probability of a destination being accepted, if it is considered
      D_j is the sum of all possible destinations, considered in order of travel cost, for zones between i and j, but excluding j,

$D_j + 1$ is the sum of all possible destinations for zones, considered in order of travel cost, between i and j, including j.

The model is calibrated by rearranging the expression for $P(D)$ to give

$$1 - P(D) = \exp(-LD)$$

Taking logarithms to the base e of both sides

$$\ln(1 - P(D)) = -LD$$

From the trip pattern the value of L can be determined by regression and L adjusted until sector to sector accuracy is achieved.

Difficulties with this method are: firstly, that L values differ for long and short trips and for this reason trips have to be stratified by trip cost (length or time); secondly, as the number of destination opportunities increase in the future, then the number of shorter trips will increase. For this reason it is usually necessary to maintain the same proportion of shorter trips in the future as are observed in the present pattern; thirdly, it is theoretically necessary to have an infinite number of destinations to utilise all trip origins. Ruiter has indicated a model revision to overcome this difficulty[9].

A comparative study of trip distribution methods carried out by Heanue and Pyers[10] suggested that the intervening opportunities model gave slightly less accurate results than the gravity model in base year simulation for Washington. The gravity model did however use socio-economic adjustment factors and without these the opportunity model was better than the unadjusted gravity model.

Blunden[11] has given the form of the competing opportunities model as

$$t_{ij} = P_i \frac{A_j}{\sum_j A_j} \bigg/ \sum_j \left(\frac{A_j}{\sum_j A_j} \right)$$

where zonal trip attractions are summed in the order of their travel cost (time or distance from the origin i.

Lawson and Dearinger[8] have used this model for the distribution of work trips in the City of Lexington and Fayette County. The model is difficult to calibrate and use and they concluded in a comparative study of trip distribution methods that the gravity model produced the best correlation with the existing trip pattern.

References

1. T. J. Fratar, Vehicular trip distributions by successive approximations, *Traff. Q.*, 8 (1954), 53–64
2. K. P. Furness, Time function iteration, *Traff. Engng Control*, 7 (1965), 458–60
3. Ministry of Transport, Land Use/Transport Studies for Smaller Towns, *Memorandum to Divisional Road Engineers and Principal Regional Planners*, London (1965)
4. J. C. Tanner, Factors affecting the amount of travel, *DSIR Road Research Technical Paper* No. 51, HMSO, London (1961)
5. S. A. Stouffer, Intervening opportunities—a theory relating mobility and distance, *Am. Soc. Rev.*, 5 (1940), 347–56

6. A. R. Tomazinia and G. Wickstrom, A new method of trip distribution in an urban area, *Highw. Res. Bd Bull.*, 374 (1962), 254-7
7. *Chicago Area Transportation Study, Final Report*, 11 (1960)
8. H. C. Lawson and J. A. Dearinger, A comparison of four work trip distribution models, *Proc. Am. Soc. Civ. Engrs*, 93 (November 1967), 1-25
9. F. R. Ruiter, Discussion on R. W. Whitaker and K. E. West. The intervening opportunities model: a theoretical consideration, *Highw. Res. Rec. 250*, Highway Research Board (1968)
10. K. E. Heanue and C. E. Pyers, A comparative evaluation of trip distribution procedures, *Highw. Res. Rec.* 114, Highway Research Board (1966)
11. W. R. Blunden, *The Land-Use/Transport System*, Pergamon, Oxford (1971)
12. Greater London Council, *Movement in London*, County Hall, London (1969)

Problems

1. Car driver work trips produced by the residents of zone 1 amount to 1000 trips. It is desired to distribute these trips to zones 1–4 which have the following characteristics.

    Zone 1 has an intrazonal time of 7 minutes and has 1000 work trips attracted to it from all zones in the study area. (The intrazonal time is the average travel time of trips which have origins and destinations within the zone.) Terminal time 2 minutes.

    Zone 2 is 15 minutes from zone 1 and has a total of 700 work trips attracted to it from all zones in the study area. Terminal time 3 minutes.

    Zone 3 is 19 minutes from zone 1 and has a total of 6000 work trips attracted to it from all zones in the study area. Terminal time 2 minutes.

    Zone 4 is 20 minutes from zone 1 and has a total of 3000 work trips attracted to it from all zones in the study area. Terminal time 3 minutes.

    The travel time factors applicable to journeys of this type are given below. The terminal time is the average time to park and walk to the destination, or walk from the origin to the car park.

Travel time (minutes)	F
1	200
5	120
7	100
11	80
14	68
16	61
17	58
18	52
20	49
21	47
23	45
25	39

2. The present and the future generated and attracted trips from four traffic zones of a transportation study are as given below, together with the present trip matrix.

Present and future generated and attracted trips

Zone	A	B	C	D
present generated trips	1500	900	1800	800
present attracted trips	1200	1000	1500	2000
future generated trips	3000	1200	2700	2400
future attracted trips	1800	3000	3500	4000

Present trip matrix

Origins	Destinations			
	A	B	C	D
A	–	400	400	300
B	200	–	300	200
C	400	300	–	600
D	200	100	300	–

(a) Calculate the first approximation to the future trips between the zones using the average factor method.

(b) Calculate the second approximation to the future trips between the zones using the Furness method.

3. Trips between the traffic zones of a proposed new town are assumed to be proportional to the trips generated by the zone of origin and the trips attracted by the zone of destination of the trip and inversely proportional to the 2nd power of the travel time between the zones. Details of three traffic zones and the value of the future trips from C to A is also given below.

Zone	Attracted trips	Generated trips
A	2400	3600
B	1600	2000
C	4000	5000

The travel time between the zones is 10 minutes.

		Zone of destination		
		A	B	C
zone of	A			X
origin	B	Y		
	C	208	Z	

What is the correct value of X, Y and Z in table 7.12?

Solutions

1	2	3	4	5	6	
Zones	A_j	Terminal times	Travel time	t_{ij}	F_{ij}	column 2 X column 6
1-1	1000	2 + 2	7	11	80	80 000
1-2	700	2 + 3	15	20	49	34 300
1-3	6000	2 + 2	19	23	45	270 000
1-4	3000	2 + 3	20	25	39	117 000

$$\Sigma\ 501\ 300$$

Using the trip distribution equation

$$T_{ij} = \frac{P_i A_i F_{ij}}{\sum\limits_{j} A_j F_{ij}}$$

where A_j is given in column 2
and F_{ij} is given in column 6

$$\text{Trips } 1\text{-}1 = \frac{1000 \times 80\,000}{501\,300} = 160$$

$$\text{Trips } 1\text{-}2 = \frac{1000 \times 34\,300}{510\,300} = 68$$

$$\text{Trips } 1\text{-}3 = \frac{1000 \times 270\,000}{501\,300} = 539$$

$$\text{Trips } 1\text{-}4 = \frac{1000 \times 117\,000}{501\,300} = 233$$

2. (a) Calculation of the trips between the zones using the average factor method.

Zone	A	B	C	D
Attraction factor	1.5	3.0	2.3	2.0
Production factor	2.0	1.3	1.5	3.0

then $\quad t'_{AB} = 400\ \dfrac{(2.0 + 3.0)}{2} = 1000$

$$t'_{AC} = 400\ \frac{(2.0 + 2.3)}{2} = 860$$

$$t'_{AD} = 300 \, \frac{(2.0 + 2.0)}{2} = 600$$

$$t'_{BA} = 200 \, \frac{(1.3 + 1.5)}{2} = 280$$

$$t'_{BC} = 300 \, \frac{(1.3 + 2.3)}{2} = 540$$

$$t'_{BD} = 200 \, \frac{(1.3 + 2.0)}{2} = 330$$

$$t'_{CA} = 400 \, \frac{(1.5 + 1.5)}{2} = 600$$

$$t'_{CB} = 300 \, \frac{(1.5 + 3.0)}{2} = 675$$

$$t'_{CD} = 600 \, \frac{(1.5 + 2.0)}{2} = 1000$$

$$t'_{DA} = 200 \, \frac{(3.0 + 1.5)}{2} = 450$$

$$t'_{DB} = 100 \, \frac{(3.0 + 3.0)}{2} = 300$$

$$t'_{DC} = 300 \, \frac{(3.0 + 2.3)}{2} = 795$$

These future trip interchanges can now be tabulated in an origin–destination matrix.

Origins	Destinations			
	A	B	C	D
A		1000	860	600
B	280		540	330
C	600	675		1000
D	450	300	795	

If complete details of the trip interchanges between all the zones of the survey had been given it would be possible to note that the future attractions and generations of the zones as obtained in this first iteration did not agree with the values given. It would then be necessary to carry out the procedure again using new values of attraction and production factors based on the ratio of first iteration interchanges to future interchanges.

(b) Calculation of the trips between the zones using the Furness method.

Iteration 1
Using the production factors calculated in (a)

	A	B	C	D
A	–	800	800	600
B	260	–	390	260
C	600	450	–	900
D	600	300	900	–
Attractions	1460	1550	2090	1760
Required future attractions	1200	2400	2300	2200
New attraction factor	0.82	1.55	1.10	1.25

Iteration 2

	A	B	C	D
A	–	1240	880	750
B	213	–	429	325
C	492	698	–	1125
D	492	475	990	–

It is stated that

$$t'_{ij} = \frac{KP_i A_j}{t^2}$$

From the data given of the trip interchange from C to A

$$208 = \frac{K \times 5000 \times 2400}{10^2}$$

$$K = \frac{208}{120\,000}$$

Then

$$X = t'_{AC} = \frac{208 \times 3600 \times 4000}{120\,000 \times 100}$$

$$= 250 \text{ trips}$$

$$Y = t'_{BA} = \frac{208 \times 2000 \times 2400}{120\,000 \times 100}$$

$$= 83 \text{ trips}$$

$$Z = t'_{CB} = \frac{208 \times 5000 \times 1600}{120\,000 \times 100}$$

$$= 139 \text{ trips}$$

8

Modal split

Trips may be made by differing methods or modes of travel and the determination of the choice of travel mode is known as modal split. In the simplest case when a small town is being considered the choice is normally between one form of public transport and the private car, with the car being used for all trips where it is available. In such a situation most trips on the public transport network are captive to public transport and very little choice is being exercised. In the larger conurbations however the effect of modal split is of very considerable significance and is greatly influenced by transport policy decisions.

Modal split should not be viewed as an entity; it is closely related in the real situation with trip generation and distribution. It has been shown in chapter 6 that additional tripmaking occurs when a private car is available and it has been observed that the destination is often influenced by the ease with which the car can be used. If the use of the private car is restricted then it is likely that the number of trips generated will be decreased rather than be made by an alternative travel mode.

Modal split may be largely determined by the availability of a particular mode of transport. In a study carried out in Riyadh by Scott Wilson Kirkpatrick and Partners it was noted that the urban form of Riyadh was characterised by a network of high-capacity expressways and freeways facilitating fast and easy access to all parts of the city. The private car was the dominant mode of transport and it was estimated that almost all people permitted to drive in Riyadh have access to a private vehicle. The study indicated that almost 97 per cent of trips were made by private transport.

In the simulation of the real system which is referred to as the transportation planning process, modal split may be carried out at the following positions in the process.

(a) Modal split may be carried out as part of trip generation whereby the number of trips made by a given mode are related to characteristics of the zone of origin. This means that transport trips are generated separately from private transport trips.
(b) Modal split may be carried out between trip generation and distribution. Car owning households in the zone of origin have a choice of travel mode depending upon the car/household ratio while non-car owning household trips are captive to public transport.
(c) Modal split may be carried out as part of the trip distribution process relating

distribution not only to travel time by mode but also with functions including the relative elasticity of demand with respect to tripmaking characteristics by the available modes.

(d) Modal split may be carried out between the trip distribution and the trip assignment process. Trip distribution allows journey times both by public and private transport to be estimated and then the modal split between public transport trips may be made on the basis of travel time and cost.

These varying approaches will now be considered in greater detail.

(a) Modal split considered as part of the trip generation process

The direct generation of public and of private transport trips was used in early work in the USA and has often been carried out in surveys of small urban areas in the UK. The Leicester transportation plan is an example of this approach where regression equations are developed for several trip types by the four modes, car driver, bus passenger, car passenger and other modes. Usually the modal split is made on the basis of car ownership in the zone of origin, distance of the zone of origin from the city centre and residential density in the zone of origin. Sometimes the relative accessibility of the zone of origin to public transport facilities is also included.

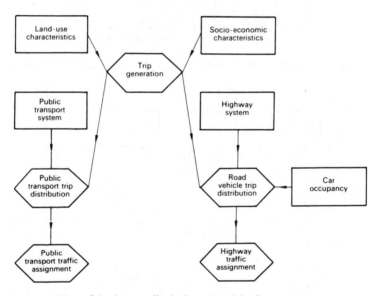

Figure 8.1   A generalised trip end modal-split procedure

This approach makes it difficult to take into account changes in the public transport network, improvements in the highway system and the restraint of private car use by economic means. Usually these models indicate a very high future car use and arbitrary modal split has to be imposed after the assignment process.

Figure 8.1 illustrates the transport planning procedure graphically when public transport and car trips are generated separately.

(b) Modal split carried out between trip generation and distribution

In this approach person trips are predicted and the percentage of these trips made
by public and private transport estimated from such factors as socio-economic and
land-use characteristics, the quality of the public transport system and the number
of cars available. The assumption is made in this method that the total number of
trips generated is independent of the mode of travel.

A typical example of this technique is given in *Transportation and Parking for
Tomorrow's Cities*[1] where the modal split decision is made entirely on the basis of
the average persons per car in the zone of origin. A generalised mapping of the
function is shown in figure 8.2. It is to be expected that socio-economic, land use
and the availability of public transport facilities would modify the form of the

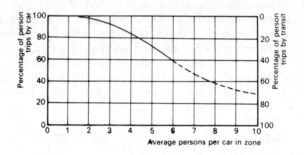

Figure 8.2   Relationship between car use and car ownership (adapted from ref. 1)

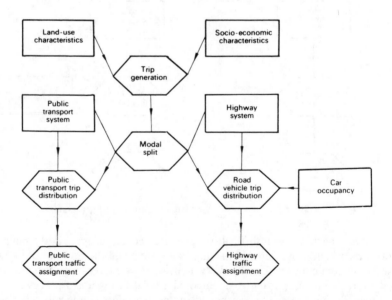

Figure 8.3   The generalised model when modal split is carried out between generation and
assignment

function and similar mappings have been used showing the relationship between percentage of trips by public transport and income.

Regression analysis allows the various factors that influence modal split to be incorporated into the analysis of existing behaviour but it is necessary not only for the regression analysis to explain the existing modal choice of tripmakers, but also to be sensitive to the factors that will affect modal choice in future projections.

In the Baltimore transportation study[1] Wilbur Smith and Associates developed the following regression equation to determine modal choice

$$T = 1.29X_1 - 1.036X_2 + 0.529X_3 + 0.150X_4 - 102$$

where T = no. of public transport trips $(R = 0.93)$
   X_1 = labour force in generating zone
   X_2 = no. of cars in generating zone
   X_3 = public transport service index
   X_4 = no. of students in zone

The transport planning procedure when modal split is carried out between generation and assignment is illustrated graphically in figure 8.3.

(c) Modal split carried out as part of the trip distribution process

An attempt to simulate human behaviour more closely and overcome difficulties inherent in the other methods is made when trip distribution and modal split are carried out at the same time for each journey made. The SELNEC study embodies this approach in which two travel modes compete at the destination end of the trip. Further details of this approach can be obtained from a study of the works of A. G. Wilson, A. F. Hawkins, D. J. Wagon and others[2,3].

(d) Modal split carried out between trip distribution and assignment

This approach is frequently used in transportation studies because it allows the cost and level of service of a trip to be used as the modal split criterion.

Because of the complexity of the transportation process, travel time alone is often used as the cost criterion. Normally travel times based on road speeds are utilised to distribute the choice trips. These travel times together with travel time by public transport are then used to determine modal split and the public transport portion of these trips is added to the captive public transport trip ends as shown in figure 8.4, which is abstracted from Movement in London[4].

An early application of modal split based on travel time is illustrated in figure 8.5 taken from Movement in London where 54 diversion curves were developed from the survey data for 3 trip purposes (work, other home-based and non-home-based), long and short trips (divided at 15 minutes of travel time by private transport excluding parking) for each of the 9 sections of the transportation study area.

Original work on modal choice has been largely carried out by Quarmby[5] using observed travel characteristics in Leeds. Quarmby developed the concept of generalised cost, a linear combination of time and money costs which he referred to as disutility of travel, and showed that a good fit to observed modal split for trips

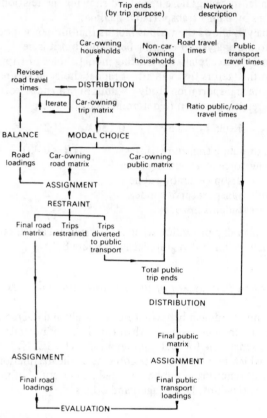

Figure 8.4 Procedure when modal split is carried out between trip distribution and assignment

between two zones could be obtained by an exponential cost difference equation of the form

$$\frac{nx}{ny} = \exp\left[-\alpha(c_x - c_y)\right]$$

where nx is the proportion of trip by mode x, ny is the proportion of trips by mode y, α is a calibration constant, and cx and cy are the generalised costs of travel between the zones by the two modes. Because nx plus ny must equal unity then,

$$nx = \frac{\exp^{-\alpha cx}}{\exp^{-\alpha cx} + \exp^{-\alpha cy}}$$

There are no particular reasons for selecting a particular modal choice procedure and the model developed will depend to a large extent on the survey information available; for this reason validation of the model against observed data is important.

The Department of Transport[6] has developed a modal choice model for new highways when a significant proportion of trips is not greater than 25 miles and

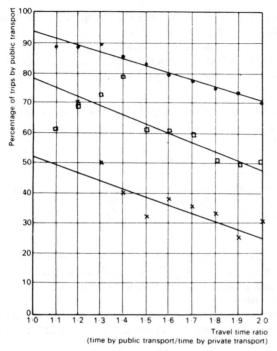

Figure 8.5 The London Traffic Survey, specimen diversion curves (adapted from ref. 4)

work trips ●
other home-based trips □
non-home-based trips ×

where the scheme is not too large and is located where the effects of the scheme are unlikely to spread over a wide area of the network.

The procedure directly determines the proportion of trips made by public transport (mst) from

$$\frac{1}{1 + \exp(Z(I_t - I_c))}$$

where I_t is a measure of impedance for public transport trips from zone i to zone j

I_c is a measure of impedance for car trips from zone i to zone j

Z is a calibration constant which varies with trip purpose.

The ratio of public transport trips to car trips (r) is also

$$\exp(- Z(I_t - I_c))$$

when this relationship is used it is found that when the impedances I_t and I_c are equal the ratio r is not equal to 1 and to allow for this an additional 'modal preference factor' d is introduced into the public transport impedance so that I_t is equal to (in-vehicle time) + (w × out-of-vehicle time) + (trip cost)/(value of time) + d. The car trip impedance is given by (in-vehicle time) + (w × car access time) + (parking cost)/(value of time) + (highway trip distance.vehicle operating cost)/(value of time) where w is a conversion factor usually taken as 2.

The factor d differs at different study locations and if previous transport studies at the location of the new scheme have not indicated a suitable value from observed travel data then the default values in table 8.1 must be used.

TABLE 8.1   Default values of Z and d (from ref. 6)

Trip purpose	Z	d	
		Central areas	Other
Home-based work	0.03	−33	+17
Home-based other	0.03	−50	+33
Non-home-based	0.04	−25	+25

To apply this method it is necessary to determine the following values for all zone to zone movements: the highway distance, public transport links and costs, vehicle operating costs, parking costs, median income and access time both to public and private transport.

An illustration of the application of this method is shown in figure 8.6 where the percentage of public transport trips as a function of cost difference is given for various trip types.

As part of the Los Angeles Regional Transportation study a modal choice model development study[7] was carried out by Voorhees and Associates. From data collected for this study the choice between public and private trips was related to the difference of disutility of using public transport and the disutility of using the private car.

This marginal utility in equivalent minutes was calculated for observed zone to zone movements using the following expression,

$$[T_r + P(T_a + T_w) + F/0.25I] - [A_r + P \times A_t + (4.76D + A_p/2)/0.25I]$$

where T_r is the public transport running time (minutes)
T_a is the access time to public transport (minutes)
T_w is the waiting or transfer time associated with public transport (minutes)
A_r is the private car running time (minutes)
F is the public transport fare (cents)
D is the highway distance (miles)
A_p is the parking cost (cents)
I is the median family income (cents per minute)
P is a psychological factor (taken as 2.5)

The value of $0.25I$ assumes that travellers value non-working travel time at 25 per cent of their wage rate.

A relationship was then developed between marginal utility and the percentage of public transport use stratified in terms of three levels of income.

The modal split procedures which only allow a decision to be made between public and private modes of transport face theoretical difficulties when a decision has to be made between more than two modes of transport. A method of determining modal split between three travel modes has been developed by Andrews and Langdon[8] and two sets of worked examples are included in this reference.

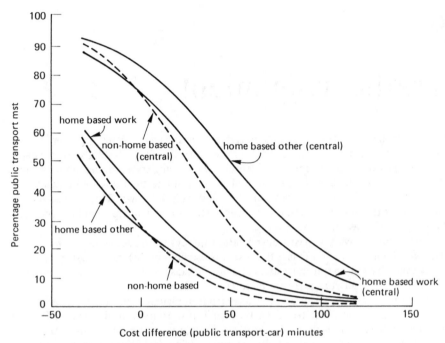

Figure 8.6   Model split for highway schemes (based on ref. 6)

References

1. Wilbur Smith and Associates, *Transportation and Parking for Tomorrow's Cities*, New Haven, Connecticut (1966)
2. A. G. Wilson, The use of entropy maximising models, *J. Transp. Econ. Policy*, **3** (1967), 108-25
3. A. G. Wilson, D. J. Wagon, E. H. E. Singer, J. S. Plant and A. F. Hawkins, The SELNEC transport model, *Urban Studies Conference* (1968)
4. Greater London Council, *Movement in London*, County Hall, London (1969)
5. D. A. Quarmby, Choice of travel mode for journey to work: some findings, *J. Transp. Econ. Policy*, **1** (1967), 273-314
6. Department of Transport, *Traffic Appraisal Manual*, London (1982)
7. A. M. Voorhees and Associates, *Los Angeles metropolitan mode choice model development study*, Los Angeles (1972)
8. R. D. Andrews and M. G. Langdon, An individual cost minimising method of determining modal split between three travel modes, *Transport and Road Research Laboratory* LR 698, Crowthorne (1976)

9

Traffic assignment

Previously the estimation of generated trip ends has been discussed together with the distribution of trips between the traffic zones. Modal split methods also have been reviewed in which the proportion of trips by the varying travel modes are determined. At this stage the number of trips and their origins and destinations are known but the actual route through the transportation system is unknown. This process of determining the links of the transportation system on which trips will be loaded is known as traffic assignment.

Apart from the very largest transportation surveys traffic assignment tends to deal with highway traffic. This is because it is usually not difficult to estimate the route taken by public transport users and also because the loading of trips on the public transport network does not materially affect the journey time.

The usual place of assignment in transportation planning synthesis is illustrated in the flow chart for Phase III of the London Transportation Study[1] (figure 8.4). Trip ends where there is no choice of travel mode, that is from non-car-owning households, are accumulated as public transport trip ends. Choice trips where a car is available are separated by the modal choice procedure into car trips and public transport trips, the public transport trips being accumulated and the car trips assigned to the network.

Usually it will then be found that the proposed road network is overloaded and some car trips will need to be restrained. If a car cannot be used then some trips will not be made at all, while other trips will be transferred to public transport and accumulated.

As the basis of assignment is usually travel time the travel times on the network links will vary the imposed loading. In addition as travel time is used in the trip distribution process it is necessary to carry out an iterative procedure between distribution, modal choice and assignment as illustrated in figure 8.4.

The change in speed with volume on a highway link is carried out using speed flow relationships for the varying highway types and it is interesting to consider just what are the effects of a speed change. Firstly it affects the choice of route because assignment is made on the basis of travel times through the network. Secondly it affects the destinations of trips because trips are distributed to varying destinations on the basis of travel time when a gravity model is used. Finally it may affect the choice of travel mode because modal choice is often made by a comparison of travel times.

There are many problems associated with speed/flow relationships, considerable variation being observed between differing highways even of the same type. There is also the additional problem that most transportation studies are based on 24 hour flows so that it is necessary to know the hourly variation and the directional distribution of flows.

In assignment it is first necessary to describe the transport network to which trips are being assigned. The network is described as a series of nodes and connecting links; in a highway network the nodes would be the junctions and the links the connecting highways. Centroids of traffic zones, at which it is assumed that all zonal trips are generated and to which they are attracted, are either at nodes or connected to them by additional links. The cost of using a link and a junction, usually in the form of travel times and delays, is given on the basis of the review of transport facilities carried out during the initial stages of the transportation survey.

There are four methods by which the assignment may be made. These are:

1. All-or-nothing assignment.
2. Assignment by the use of diversion curves.
3. Capacity restrained assignment.
4. Multipath proportional assignment.

All-or-nothing assignment

In this method an algorithm is used to compute the route of least cost, usually based on travel time between all the zone centroids. For each zone centroid selected as origin, a set of shortest routes from the origin to all the other zone centroids is referred to as a minimum tree. When the trips between two zones are assigned to the minimum path between the zones, then the assignment is said to take place on an all-or-nothing basis.

There are obvious difficulties with such a simplified approach, some of which are inherent in the other assignment methods. It is obviously incorrect to assume that all trips commence and terminate at a zone centroid. If the length of the links within the zones is small compared with the length of remainder of the minimum link path, then the errors may not be so serious. Because of its simplicity, travel time is usually employed as a measure of link impedance, but travel times may not be precisely estimated by the traveller. The use of a cost function which reflects the perceived cost of travel is desirable. The loading on a link in this method is extremely sensitive to estimated link and node costs. If these have been incorrectly estimated, then the resulting assignment is open to question. There is also the problem that links with small travel costs will attract trips without any adjustment in link cost.

Assignment by the use of diversion curves

Originally diversion curves were used to estimate the traffic that would be attracted by a single new route or transport facility. It is thus necessary to compare travel cost with and without the new transport facility, the decision as to whether a trip would use the new facility being based on a cost ratio or difference between the with and the without situation. When diversion curves are used for assignment it is necessary to consider the network with and without the new facility or highway link as input to an all-or-nothing assignment. The travel costs from the two networks are then

used with the diversion curve to determine the proportion of the trips that are
diverted from the existing network and transferred to the network with the new
facility.

An early application of this technique was used in the traffic studies for the first
section of the M1 motorway[2] where 50 per cent of trips were diverted to the motorway if the cost (time) of the trip on the existing route was within ±10 minutes of
the trip cost on the motorway. It has often been considered that differences of trip
cost more effectively modelled human behaviour than ratios. A diversion curve of
this type is illustrated in figure 9.1.

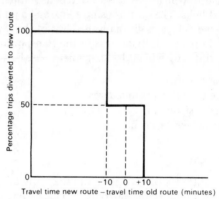

Figure 9.1   Early diversion curve

It is now usual to employ diversion curves similar to one shown in figure 9.2,
which has been derived from observation of driver behaviour[3].

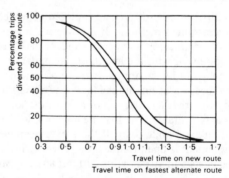

Figure 9.2   Specimen highway diversion curves (adapted from ref. 3)

In the USA indifference curves have been developed that take into account both
time and distance savings. Figure 9.3 shows the form of the relationship proposed
by Moskowitz[4] and based on observations of traffic diversion to new highways.

The mathematical form of the relationship is

$$\text{Percentage diverted to new route} = 50 + \frac{50(d + 0.5t)}{\sqrt{[(d - 0.5t)^2 + 4.5]}} \tag{9.1}$$

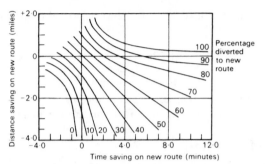

Figure 9.3 Highway diversion curves based on time and distance savings (adapted from ref. 4)

where d = distance saving on new route

   t = time saving on new route.

Drivers perceive costs but as they are able to measure speed and form an assessment of journey speed, speed can be used to assess the diversion of drivers to a new route. This may well be true of the present motorway system in this country in that drivers are reasonably certain of being able to maintain a given speed but for certain journeys may not be aware if their journey time is greater on the new route. The diversion relationship shown in figure 9.4 developed for the Detroit Area Traffic Survey[5] illustrates the form of the assignment curves.

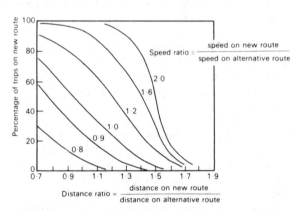

Figure 9.4 Highway diversion curves based on distance and speed ratios (adapted from ref. 5)

Several mathematical forms have been suggested for diversion curves and empirical curves can be derived. The Traffic Assignment Manual[6] gives an example of a diversion curve derived from the Burrell method where the proportion of traffic P_2 on route 2 is given by

$$P_2 = f(x)$$

where $f(x)$ is the normal integral

   x is $\dfrac{(\bar{C}_1 - \bar{C}_2)}{I(\bar{C}_1 + \bar{C}_2)}$

$\bar{C}_1$ is the mean disutility on route 1 (usually travel time or generalised travel cost)

$\bar{C}_2$ is the mean disutility on route 2

I is $V^2/3$

and V is the Burrell variation term as a proportion of unity where V^2 has been found equal to 0.706 minutes from empirical evidence

Capacity restrained assignment

All-or-nothing assignment results in a link with a favourable cost attracting a considerable number of trips, while links with unfavourable costs attract few trips. In practice this would result in the originally favourable link becoming overloaded, a situation which would not occur in real life.

The problem usually occurs when assigning trips to a highway network because in practice a balance exists between travel costs and flows. A number of routes between an origin and a destination are selected by tripmakers so that the perceived travel cost is approximately equal on all the routes. With a public transport network the choice of routes is more limited and the major difficulties occur when incorporating walking, waiting and interchange times.

Capacity-restrained assignment to highway networks attempts to tackle this problem by taking into account the relationship which exists between speed and flow on a highway. These relationships are discussed fully when considering the evaluation of transportation proposals.

The Department of Transport[8] recommends the use of capacity restrained assignment when considering major road schemes when traffic congestion dissuades a significant number of drivers from using otherwise attractive routes. Because of the extra resources required they suggest the use of capacity restraint only when all or nothing or multi-routing would not adequately spread the traffic over the network.

There are several basic approaches to this type of assignment which have been used for capacity restrained assignment programs. The Bureau of Public Roads Model[7] first makes a complete all-or-nothing assignment to the network. The journey cost, usually expressed as time, is then updated on the basis of the flow assigned to it, and the whole procedure repeated for several iterations until the link costs show only a limited change on each iteration.

An alternative non-iterative method used in the Chicago Area Transportation Study required the selection of a random origin zone, the calculation of the minimum tree and the assignment of flows from this zone to the tree. Journey costs on the links were then revised using the actual link flows and the procedure repeated with another randomly selected zone of origin.

Multipath proportional assignment

In urban areas there are many alternative routes between a given origin and destination and in actual fact tripmakers would be distributed over all these routes. This is because tripmakers will be unable to judge the route of least cost accurately and different tripmakers will make differing decisions. Multipath proportional assignment attempts to simulate this situation by assigning proportions of the trips between any two zones to a number of alternative routes.

Burrell[1] has described a model in which it is assumed that the tripmaker does not know the actual cost of using a link but that he associates a supposed cost with each link. This supposed cost is randomly generated from a distribution with the actual link cost as its mean and a given standard deviation representing the inability of a tripmake, to judge costs effectively. The assignment for one zone is then made on an all-or-nothing basis using the randomly determined link costs. Assignment of trips from the next zone considered will then be made on the basis of a further set of randomly determined link costs and the process repeated until all trips have been assigned to the network.

It has been reported that the assignment made by this model produces a satisfactory comparison with actual network flows giving more satisfactory results than all-or-nothing assignment.

Public transport assignment

Public transport assignment differs from the assignment of highway trips in that the public transport system is composed of a system of fixed services following stated routes at regular intervals. Because of the greater complexity of the assignment process the trips are usually assigned on an all-or-nothing basis. In the case of the public transport network the journey times that form the basis of the assignment must include walking, waiting and link travel times in addition to waiting time caused by the necessity to change from one route to another.

An assignment program developed for public transport and used for the London Transportation Study is Transitnet[10] developed by Freeman, Fox, Wilbur Smith and Associates. The program describes the network by routes. Each route has a frequency, modes at which a transfer to either a walk link or a centroid connector can take place and a travel time between nodes. Waiting time values that are functions of the frequencies of the routes are used and there are time penalties built in to model a traveller's reluctance to change route or mode.

References

1. Greater London Council, *Movement in London*, County Hall, London (1969)
2. Road Research Laboratory, The London–Birmingham Motorway—traffic and economics, *Tech. Pap. Rd Res. Bd* 46, HMSO, London (1961)
3. R. E. Schmidt and M. E. Campbell, *Highway traffic estimation*, Eno Foundation for Highway Traffic Control, Saugatuck, Connecticut (1956)
4. K. Moskowitz, California method of assigning diverted traffic to proposed freeways, *Highw. Res. Bd Bull*. 130 (1956)
5. *Detroit Metropolitan Area Traffic Study*, Detroit, Michigan (1956)
6. Bureau of Public Roads, *Traffic Assignment Manual*, Washington DC (1964)
7. J. D. Carroll, A method of traffic assignment to an urban network, *Highw. Res. Bd Bull*. 224 (1959)
8. Department of Transport, *Traffic Appraisal Manual*, London (1982)
9. J. E. Burrell, Multiple Road Assignment and its application to capacity restraint. Proceedings of the 4th Internal Symposium on the Theory of Traffic Flow, *Strassenbahn und SStrassenverkehrstechnik* 86 (1969)
10. R. B. Hewing and M. L. H. Hoffman, An explanation of the Transitnet algorithm, *Research Memorandum* 198, G.L.C. Department of Highways and Transportation (1969)

10

The evaluation of transportation proposals

The object of the simulation of the land-use/transportation process is to estimate the trips that will be attracted to a proposed future transportation system for a given pattern of land-use development. To compare effectively different transportation proposals it is imperative that each proposal should be evaluated.

There are several grounds on which the end produce of the simulation process must be judged. These are tabulated below:

1. It is necessary to be certain that the numerical results output by the computer are realistic and all the computer programs are functioning correctly.
2. It is necessary to check that the predicted future transportation requirements can be met by the proposed transportation system being evaluated.
3. It is necessary to estimate the economic consequences of the provision and operation of the system under evaluation.
4. It is necessary to consider the environmental effects of the operation of the proposed transportation system.

These sections will now be considered in greater detail.

Program checking to ensure that numerically correct results in accordance with the mathematical model are incorporated into the program is a very tedious process and even after experience with transportation programs doubt can sometimes be felt about the validity of the results obtained. A frequent source of error is incorrect input information and this should be carefully checked to ensure that the input data is correct.

The second stage in the evaluation of a transportation plan is to ensure that the proposed system is capable of dealing with the flows assigned to it by the simulation process. Many transportation simulation programs deal with 24-hour flows while critical flows usually occur in practice during the peak hours. It may be possible to estimate the peak hour flows from the 24-hour flows or alternatively the transportation programs may be run to deal with a peak hour flow.

Where highway networks are being evaluated it may be found that where there is no restriction on modal choice the highway system may be grossly overloaded. This problem may be overcome by the use of modal choice relationship that incorporates a travel time by public transport ratio or by the arbitrary transfer of trips from the private transport to the public system.

Finally it should be remembered that the predictions of future trips on the transport system are dependent on the initial assumptions made. Where all-or-nothing assignment is used, for example, the network may have a section of highway where heavy traffic flows are predicted while an adjacent link, which has a longer travel time, may be lightly loaded. In practice the trips would be distributed between the adjacent links and predicted flows should be considered on the basis of corridors of movement.

In the past the environmental effects of proposed transport systems received less consideration than the effects of capacity or economic considerations. It is unfair to criticise transportation planners for ignoring environmental considerations in the past. Normally they were required to provide the least-cost solution and since damage to the environment had in the past not been given a high economic price, solutions that damaged the environment were preferred.

There are however considerable problems in placing a value on environmental factors and this has been partially responsible for the failure to consider them. The most important environmental effect at the present time is that of noise and whilst it is possible to predict the noise levels that will be caused by major highway proposals, the valuation of the effects of noise is less easy to make.

Transport facilities usually have an adverse effect on the environment: they cause noise and vibration, highway traffic pollutes the air and frequently intrudes both physically and visually. The disturbance to the environment must however be balanced against the result of doing nothing. A situation in which traffic is congested or there is a high accident risk may damage the environment to a greater extent than a new highway or other transport system.

The cost of reducing the damage to the environment caused by a new transport system may however be better spent providing alternative facilities that may improve the environment. This factor should be seriously considered at a time when any new transport proposal arouses intense opposition from those immediately affected.

Economic evaluation of transport schemes

It is however the economic consequences of carrying out transportation schemes which have received increasing attention during the last decade and it is becoming increasingly necessary to justify any new proposal on economic grounds.

This evaluation of a transport scheme on economic grounds is referred to as a cost-benefit analysis. As would be expected, the costs of carrying out a scheme are compared with the benefits that may be anticipated when the scheme is completed and in operation.

The usual method adopted in the United Kingdom for the economic evaluation of highway schemes when the benefits of the scheme occur over a considerable period of time is to discount the costs and benefits of the scheme so that they represent present day values. In this way all the costs and benefits are converted to a net present value. Any scheme which has a positive net present value is economically worthwhile and when similar schemes are being compared on economic grounds, the scheme which has the greatest net present value is preferred.

Discounting is the reduction in value of future costs and benefits by dividing future values by $(1 + r)^n$ where r is the discount rate raised to the power n, the number of years hence when the cost or benefit is incurred. This means that when the discount rate is taken as 10 per cent and a benefit of £1000 will be obtained in 5 years' time, then the present value of this benefit is £621. If discounted benefits are considered positive and discounted costs are considered negative, then their sum over the period of time being considered is the net present value of the scheme.

For example, consider a proposal to upgrade the signal control at a highway junction at cost of £15 000 and with associated future maintenance costs as shown in table 10.1. The benefits of upgrading the signal control will be a reduction in delay to vehicles passing through the junction and when this is valued it is estimated it produces the benefits also shown in table 10.1.

In table 10.1 the costs and benefits of the scheme are tabulated for a 10 year period; the discount rate selected is 7 per cent and the sum of the discounted net benefits is calculated as £42 747. As this is positive, the scheme is economically viable.

It is of course possible that with periodic maintenance the improvements to the signal control will have a life considerably longer than 10 years. It is usual in the United Kingdom to consider costs and benefits over a period of 30 years even though many highway works will have a residual value greater than this period of

TABLE 10.1   Calculation of net present value

Year of operation	Capital and maintenance cost (£)	Value of annual reduction in delay (£)	Net benefit (£)	Discounted net benefit (%) (£)
1	25 000	1 500	−23 500	−21 963
2	2 000	3 000	1 000	873
3	2 000	5 000	3 000	2 449
4	4 000	9 000	5 000	3 815
5	2 000	13 000	11 000	7 843
6	2 000	16 000	14 000	9 329
7	2 000	19 000	17 000	10 587
8	4 000	20 000	16 000	9 312
9	2 000	21 000	19 000	10 335
10	2 000	22 000	20 000	10 167
				42 747

time. It is considered unimportant to consider costs and benefits beyond 30 years because of the uncertainty associated with long-term forecasts and also because the present value of a £1 cost or benefit in 30 years time is only 13p when discounted at 7 per cent.

There are usually several alternative schemes which can be proposed to alleviate transport problems, each with differing costs and associated benefits. When a choice is made purely on economic considerations, then the scheme with the highest net present value would normally be preferred.

Cost–benefit analysis applied to United Kingdom trunk road investments

Cost–benefit analysis was first used in the assessment of trunk road schemes in the United Kingdom in the 1960s; later, in the 1970s, the Department of Transport developed the computer cost–benefit analysis program COBA[1] where the benefits derived by road users are calculated and expressed in monetary terms.

When a scheme does not have a marketable output, as is the case when users are not charged directly for road travel, then cost–benefit analysis may be applied. It is not possible to value all the costs and benefits of a highway scheme and in most instances the benefits of a highway project have been limited to time saving benefits, savings in operating costs, and savings in accident costs. These benefits are related to capital and maintenance costs. As has been previously discussed, costs and benefits occur over a period of time and to relate them to the present time, discounting is used. In the valuation of costs and benefits, it is resource costs which should be used with taxes and subsidies, which are merely transfer payments, eliminated.

The first stage in the COBA appraisal program is to determine the alternative schemes which are to be subjected to cost–benefit analysis. Because of the pressures of public opinion there will normally be several alternatives for any road proposal, the minimum number is of course two: the 'Do-Minimum' and the 'Do-Something'.

The 'Do-Minimum' option is frequently the situation where road improvements are not to be carried out and traffic will have to travel along existing facilities. There are many cases however when some improvements will be carried out regardless of whether the scheme being appraised is constructed. An example could be where improvements to a signalised junction will provide better pedestrian facilities and will be carried out regardless of major road improvements. A further example might be where a relatively small highway improvement or low-cost traffic management would considerably improve traffic conditions; in this case the 'Do-nothing' option would offer an unrealistic alternative to the scheme being assessed.

When a highway improvement is carried out it will result in changes in the trip pattern. Existing traffic may take a new route; that is, re-assignment of the trip pattern will take place. Existing traffic may change its destination because of the convenience of the new road; that is, re-distribution may take place. New traffic may be attracted on to the new system because of the existence of the new road; that is, generation of traffic may take place. Trips may be attracted on to the new road which previously went by another mode of travel; that is, a modal split. Lastly, trips may be made at another time of day compared with the 'Do-Minimum' situation.

Because COBA deals with highway schemes whose effects may be largely restricted to a corridor of travel, only the effect of redistribution is considered. Where the road scheme is in a congested urban area, or if a complete long inter-urban route or an estuarial crossing is being appraised, then the other factors may have an important effect on the resulting traffic pattern.

The major benefit arising from the construction of a road scheme is reduction in travel time for vehicles travelling along links and through junctions. These time savings have to be assigned a monetary value so that they may be compared with construction and maintenance costs. The value of a time saving depends upon the purpose for which the trip was being made. The COBA program recognises two trip purpose and values time as 'working time' and 'non-working time'. As many trips in peak hours are for travel to and from work, this is often an important element of time saving; in the COBA program it is classified as 'non-working time'.

Working time is valued as the cost to an employer of the travelling employee; it consists of the gross wage rate together with overheads. Variations in wage rates among the population of car drivers, car and bus passengers are taken into account by the use of a mileage-weighted average income taken from national statistics. The value of 'non-working time' has been derived from studies of the choice made by travellers between a slower cheaper mode of travel or a faster more expensive mode. These studies indicate that on average the value of in-vehicle non-working time is equivalent to 25 per cent of the hourly wage rate.

It has frequently been argued that small time savings do not have the same unit value as large time savings resulting from major road schemes. In the COBA program all time savings have the same unit regardless of their duration.

The COBA program calculates time user costs for the 'Do-Minimum' and the 'Do-Something' options by disaggregating the traffic flows by vehicle category and by flow group; the flows are then further divided into work and non-work journeys. The time cost per vehicle is a function of vehicle occupancy which varies with vehicle type and journey purpose. The hourly flow of vehicles is then converted into a flow of people whose time can be valued to produce time costs for links and junctions.

Savings in vehicle operating cost between the 'Do-Something' and the 'Do-Nothing' options are one of the benefits of road improvement schemes. They depend on link distance and on link speed; usually they are positive but depending upon the relative changes in distance and speed may possibly be negative. In the COBA program, vehicle operating cost is related to distance travelled and is a function of fuel, oil, tyres, maintenance and depreciation and the size of vehicle fleets. The resource cost of fuel consumption depends on average link speed and the hilliness of the link; the relationship gives a high cost at low speeds as would be expected with stop-start conditions.

The marginal resource costs of oil and tyres are a fixed cost per kilometre, maintenance is in part related to link speed and in part varies with speed.

Depreciation for vehicles, other than cars, is entirely related to mileage; for cars it is related to the passage of time as well as to mileage; only the latter is included in car vehicle operating cost. Increased link speeds may result in a reduction in the number of working cars, commercial vehicles and public service vehicles which operators need to provide; this is taken into account in vehicle operating cost.

An important element in the benefits which arise from road schemes is the change in the number and severity of accidents consequent on the change from the 'Do-Minimum' to the 'Do-Something' option. To allow accident reduction to be considered in the COBA program it is necessary to place a monetary value on accidents. It is assumed that the cost of an accident is composed of three elements. Firstly, the direct costs to individuals and organisations involved in the accident, that is, vehicle damage, police and medical costs. Secondly, the lost output of those killed or injured measured by the expected loss of earnings together with non-wage payments such as insurance contributions paid by an employer. Thirdly, an allowance is made for the 'pain, grief and suffering' which results from injury or death in a road accident; its value in COBA is stated to be largely notional and minimal.

Accident severity has been found to vary with road type because of differences in the number of casualties per accident and in the amount of damage per accident. As would be expected an accident on a rural road will result in more casualties per accident and greater damage than on an urban road where speeds may be expected to be lower.

With the monetary value of accidents having been established, it is necessary to know the accident rate for the particular road types being considered in the 'Do-Minimum' and the 'Do-Something' situations. The only complete record of road accidents in the United Kingdom are those where personal injury takes place, and for this reason accident rates in COBA are expressed as personal injury accidents per million vehicle kilometres and an allowance made in the valuation of these accidents for damage only accidents. It is preferable to separate accidents which take place on links between junctions and those which take place at or adjacent to the junctions. For accidents at junctions, COBA uses relationships between junction flows and accidents for various junction types. When the 'Do-Minimum' option is being considered and little change is contemplated, then this option become the 'Do-Nothing' option and existing accident data is preferred. For the 'Do-Something' option then, predicted values based on past accident experience on similar road types must be used.

The benefits of the road scheme in the COBA program are taken as the monetary value of the reduction in travel time and accidents and usually the reduction in vehicle operating cost, and these are compared with construction cost, delays to traffic during construction and the changes in maintenance cost due to the construction of the new road. The comparison between costs and benefits is made for a 30 year period on the basis of their present value.

The effects of re-distribution, generation and modal split on economic assessment of road schemes

As has been previously stated, the COBA program only considers the redistribution of trips within the highway network assuming that the reduction in travel cost will not cause any change in the demand for travel. In many cases this will be incorrect and additional trips will be generated and a change in travel mode could take place if the scheme is of considerable magnitude. In addition, any increase in traffic flow will lead to a reduction in journey speed, an increase in travel time and an increased trip cost. These effects on the trip matrix are discussed in chapter 27. When economic assessment of a road scheme where these effects are considered important is carried out, complexities arise because car travellers usually perceive the cost of a trip to be less than the total user cost, normally only considering fuel costs which include taxation costs but which are excluded in cost-benefit analysis as being a transfer payment. The car traveller frequently ignores non-fuel costs and these have to be included in an economic assessment.

Forecasting future vehicle ownership and use

In the assessment of road schemes it is necessary to take into account the increase in the number of vehicles which can be expected in the future. In particular it is the growth in the number and use of private cars which has caused problems to the transport planner and the highway engineer working in urban areas. The rapid growth in the number of private cars in the United Kingdom during the 1950s and 1960s led to research on the prediction of vehicle ownership and use and in particular car ownership and use.

The earliest methods of forecasting car ownership were based on logistic growth curves of cars per head. In 1965 Tanner[2] postulated that, except for tractors, the forecasts of vehicles per head should lie on a logistic curve determined by the 1964 levels of ownership and future saturation levels of ownership when further increases in car ownership per person ceased.

When the logistic relationship is assumed, when the percentage increase per year is plotted as the ordinate and the corresponding cars per head plotted as abscissa then a linear relationship is observed and the value of car ownership when the percentage increase per year is zero is the saturation level. Using data available at that time, Tanner estimated that the saturation level in Great Britain would be 0.45 cars per head with an alternative estimate of 0.40 if a moderately restrictive attitude to motoring was taken. With the logistic relationship defined, car ownership could be determined for any year in the future and when multiplied by population estimates produced future levels of car ownership. Car traffic was considered to increase with car ownership as it was assumed that average annual car travel would remain constant.

Such a simple relationship was criticised in that it took no account of economic or social circumstances. As a consequence and shortly after the economic changes of 1973, Tanner presented a revised model of car ownership[3] in Transport and Road Research Laboratory Report LR 650 which although still adopting a logistic relationship introduced the economic variables gross domestic product and fuel prices. At the same time the kilometres per car per year were assumed to vary with gross domestic product and the length of the motorway network. This report introduced variability into forecasting by assuming gross domestic product growth rates of 2, 3 and 4 per cent and saturation car ownership levels of 0.42, 0.43 and 0.44 cars per head.

In response to continued criticism that car ownership and car travel did not fully take into account the underlying causes of growth and travel, Tanner proposed in 1977[4] a power growth model where the second half of growth takes longer to develop than the first half of the curve. The factors which determined growth in this model were the passage of time, income in terms of gross domestic product per person, and the cost of motoring. A range of saturation levels were used: 0.44, 0.46 and 0.48; gross domestic product growth rates of 2, 3 and 4 per cent; and fuel price rises of 4, 2 and 0 per cent.

At this time there was a growing interest in models which used the factors which predicted car ownership and travel rather than models which depended on the passage of time. A casual model of this form was proposed in Bates[5] in 1978 where models for the proportions of households with at least 1 car and for households with at least 2 cars were proposed in terms of the ratio of household income to the index of car purchase costs, the latter acting as an income deflator. This model was developed from information in the Family Expenditure Survey of Great Britain and the relationship was found to be reasonably constant over the period 1965-75. In the period 1975-78 however car prices rose more rapidly than real gross household income and the model indicated that car ownership would fall but in fact car ownership continued to increase and the model under-estimated the 1978 level by 16 per cent.

Because of difficulties with the Bates model the Department of Transport in their National Road Traffic Forecasts, 1984[6] adopted a model which related the probability of ownership of at least 1 car per household and at least 2 cars per household to income deflated by the retail price index and also by the number of driving licences per adult. When making the 1984 forecasts of road traffic the Department of Transport also considered the forecasts of the LR799 model but using additional data were able to obtain new estimates of the relationship between car ownership, income and motoring costs based on time series data. The new estimate of income elasticity is in a range of 0.15 to 0.18 (previously assumed to be 0.35 to 0.41) and the new estimate of elasticity with respect to motoring costs is in the range -0.13 to -0.17 (previously -0.22 to -0.27) indicating that the effect of time is more important than was previously assumed.

The environmental evaluation of transport schemes

Major transport proposals require much more than an engineering and economic appraisal and it is well established that consideration needs to be given to the environmental effects of projects. In addition, the increasing necessity for public consultation during the initial planning stage of transport schemes, and the emphasis placed on environmental issues by protest groups, have made the assessment of environmental planning increasingly important.

Environmental issues became important in the United States during the 1960s and a procedure suggested by Leopold et al.[7] has been modified for use in the United States.

Leopold stated that an assessment of the probable impacts of the variety of the specific aspects of the proposed action upon the variety of existing environmental elements and factors should consist of three basic elements:

(a) a listing of the effects on the environment which would be caused by the proposed development and an estimate of the magnitude of each;
(b) an evaluation of the importance of each of these effects;
(c) the combining of magnitude and importance estimates in terms of a summary evaluation.

For each scheme being assessed these three elements are analysed using a matrix, on one axis of which are the actions which cause environmental impact and on the other axis are existing environmental conditions which might be affected.

There are a total of 100 environmental actions which may cause environmental impact, grouped into modifications of the regime, land transformation and construction, resource extraction, processing, land alteration, resource renewal, changes in traffic, waste emplacement and treatment, chemical treatment, and accidents. Listed vertically are 88 environmental characteristics grouped into physical and chemical characteristics, biological conditions, cultural factors, and ecological relationships.

For each action which is expected to have a significant interaction with an environmental condition the relevant square in the matrix is divided by a diagonal which runs from the upper right to the lower left.

After all anticipated actions have been checked with possible environmental effects, a decision is made as to the magnitude and also the importance of the interaction. Within each box, representing a significant interaction, a weighting factor is introduced ranging from 1 to 10. The number is placed in the upper left-hand corner to indicate the relative magnitude of the interaction and in the lower right-hand corner to indicate the relative importance of the interaction. A value of 1 indicates the least magnitude or importance of the interaction and a value of 10 the greatest magnitude or importance.

For an example, the circular considers the case of an engineering proposal which requires the construction of a highway and a bridge. The action will have environmental effects which may be classified under erosion, deposition and sedimentation. It may be that, because of poor consolidation of the soil in the region of the bridge, erosion is likely to be considerable and so the magnitude of the impact will be weighted with a factor of 6 or more. If, however, the river already carried large sediment loads and further erosion will not have undesirable effects, then the weighting of the importance of the interaction may be 2 or less.

If possible, assignment of numerical weights to the magnitude and importance of an interaction should be based on factual data. When all significant interactions have been considered then a simplified matrix can be presented containing the product of the magnitude and importance weightings, beneficial interaction products being shown as positive and detrimental interaction products shown as negative values.

Frameworks for road appraisal

In the United Kingdom it is usual to represent the effect of major road construction proposals by the use of a framework. This is simply the tabular presentation of data summarising the main likely direct and indirect impacts on people of the alternative options for a proposed highway scheme[8]. For trunk road scheme appraisal the Department of Transport state that each option for the construction of the road, which is technically feasible and considered to be an acceptable and clearly separate solution, will have a separate column in the framework. The minimum number of columns will be one containing details of the officially proposed route, one containing the forecast changes if a new route is not constructed (that is, the 'Do-Nothing' or the 'Do-Minimum' solution), and finally a column for comments.

For the data presented in the framework to be consistent from scheme to scheme it is recommended that information be presented in the following groups: the effects on users of facilities, the effects on policies for conserving and enhancing the area, the effects on policies for development and transport, and finally financial effects.

In the first group are the effects on travellers; defined as vehicle drivers and passengers, cyclists and pedestrians; in terms of time savings or delays, changes in vehicle operating costs, accident reductions, driver stress, view from the road, and amenity and severance.

In the second group, the effects on occupiers of property, occupiers are sub-divided into occupiers of residential property, industrial and commercial property, schools and hospitals, public or special buildings, recreational space, and agricultural land. The impacts on occupiers which should be considered are: number affected by demolition, changes in noise level, visual effects, severance, disruption during construction and landtake.

In the third group, users of facilities are sub-divided into users of: shopping centres, public buildings such as churches, libraries or community centres, and users of recreational areas and facilities.

The impacts described in the fourth and fifth group are the formal views of the highway authority themselves. Whilst in group six the economic costs and benefits are presented for each option in terms of the net present value.

The impact of traffic noise

Noise levels generated by highway traffic can be measured or calculated and the subject is considered in detail in chapter 25. It is however the reaction of human beings to noise levels which is of importance in attempting to determine the impact of the noise. Different people have different reactions to the same noise level and it is necessary to determine the distribution of responses to noise by the use of attitude surveys.

Social surveys have been carried out in which respondents were asked to give their reactions to traffic noise levels experienced at home. Using a seven-point scale, they were asked to express their reactions. A score of 1 was labelled 'definitely satisfactory' and a score of 7 was labelled 'definitely unsatisfactory'.

It was found[9] that a family of curves could be derived, of the form

$$p_s = a + \frac{1}{(b + e^{-c(L-d)})}$$

where p_s is the proportion of people with scale score $\geq S$ at noise level L
a, b, c, d are constants depending upon the value of S
L is the 18 hour L_{10} noise level (dB(A))

Score   Equation for proportion scoring as shown

$S > 1$ $p_1 = 1.0$

$S > 2$ $p_2 = 0.2 + \dfrac{1}{(1.25 + \exp - 0.104(L - 56))}$

$S > 3$ $p_3 = 0.1 + \dfrac{1}{(1.11 + \exp - 0.116(L - 63))}$

$$S > 4 \qquad p_4 = 0.08 + \frac{1}{(1.09 + \exp - 0.129\,(L - 68))}$$

$$S > 5 \qquad p_5 = 0.06 + \frac{1}{(1.06 + \exp - 0.144(L - 73))}$$

$$S > 6 \qquad p_6 = 0.02 + \frac{1}{(1.02 + \exp - 0.113(L - 78))}$$

$$S > 7 \qquad p_7 = 0.0 + \frac{1}{(1.0 + \exp - 0.096(L - 86))}$$

Using these equations, the proportions in various scale score ranges can be generated. If the numbers of people within the noise interval of which L is the mid-point is known, then the numbers with scale scores $> S$ at L can be calculated. To obtain the number of people with scale score n, the number with a scale score $\geqslant n + 1$ is subtracted from the number with scale score $\geqslant n$.

To aggregate these scale scores over an area subject to highway noise impact it is necessary to make a decision as to how they should be weighted. It is suggested that a scale score of 2 ('not at all bothered by the noise intrusion') represents a neutral attitude and that a progressively less favourable attitude is indicated by higher scores. The intensity-weighted scale score ranges from +1 (basic scale score = 1) to −5 (basic scale score = 7). These weightings allow the weighted estimation of the number of people affected by a proposed highway scheme to be determined.

It is necessary to combine noise impacts for different land uses where these vary. It is suggested that the following procedure be used.

(1) Where the land use is such that speech communication is important (such as schools), 5 dB(A) should be added to the anticipated noise level and the distribution of scale scores for residential use should be applied to this modified noise level, together with a duration weighting (see below).
(2) Similarly, where a low level of noise is desirable (such as religious buildings, parkland), 10 dB(A) should be added to the anticipated noise level and a duration weighting applied.
(3) When considering pedestrians, 10 dB(A) is added to the anticipated noise level, because pedestrians receive a noise impact which is not attenuated by a building structure. As before, a duration weighting is applied.

Duration weightings for length of exposure to noise in various situations are as follows:

Residential land use	1.0
Educational establishment	0.5
Hospitals	1.5
Churches	0.2
Parks	0.2
Pedestrians	0.1

A measure of the noise impact of a highway proposal may thus be obtained by summing the products of the people affected by noise impact, their intensity weighted scale score and the appropriate duration weight.

As an example of the above procedure consider a proposed highway scheme which results in a residential area containing 600 people being submitted to an 18 hour L_{10} noise level in the range 64-66 dB(A). In a school 300 children and staff will be subjected to an 18 hour L_{10} noise level in the range 62-64 dB(A) and a church with an average attendance of 200 people will be subjected to an 18 hour L_{10} noise level in the range 64-66 dB(A). Calculate the intensity and duration weighted number of people exposed to noise impact. Considering the residential development where the 18 hour L_{10} noise level is in the range 64-66 dB(A).

The proportion of people with a scale score > 1 is 1.0

The proportion of people with a scale score > 2

is $0.2 + \dfrac{1}{(1.25 + \exp - 0.104(L - 56))}$

$0.2 + \dfrac{1}{(1.25 + \exp - 0.104(65 - 56))}$

$=$ 0.81

The proportion of people with a scale score > 3

is $0.1 + \dfrac{1}{(1.11 + \exp - 0.116(L - 63))}$

$0.1 + \dfrac{1}{(1.11 + \exp - 0.116(65 - 56))}$

$=$ 0.63

The proportion of people with a scale score > 4

is $0.08 + \dfrac{1}{(1.09 + \exp - 0.129 (L - 68))}$

$0.08 + \dfrac{1}{(1.09 + \exp - 0.129 (65 - 68))}$

$=$ 0.47

The proportion of people with a scale score > 5

is $0.06 + \dfrac{1}{(1.06 + \exp - 0.144(L - 73))}$

$0.06 + \dfrac{1}{(1.06 + \exp - 0.144(65 - 73))}$

$=$ 0.30

The proportion of people with a scale score > 6

is $0.02 + \dfrac{1}{(1.02 + \exp - 0.113(L - 78))}$

$$0.02 + \frac{1}{(1.02 + \exp - 0.113(65 - 78))}$$

$= \quad 0.21$

The proportion of people with a scale score > 7

is $\quad 0 + \dfrac{1}{(1.0 + \exp - 0.096(L - 86))}$

$$\frac{1}{(1.0 + \exp - 0.096(65 - 86))}$$

$= \quad 0.12$

The proportions with scale scores equal to $1, 2, 3, 4, 5, 6$ and 7 are calculated by subtraction and are entered in table 10.2. In a similar manner the proportions of people in the school and the church are calculated and also placed in table 10.2. The intensity and the duration of the weightings are also placed in the table and finally by multiplication and by addition the number of people affected weighted by intensity and duration is obtained. Similar calculations can be made for other highway alignments and a comparison made between the noise impact of the options which are being evaluated.

A simpler approach is suggested by the Department of Transport in the Manual of Environmental Appraisal where it is suggested that in the preliminary stages of scheme preparation it is sufficient to list the number of properties that are wholly or partly within distance bands of 0-49, 50-99, 100-199 and 200-300 m from the road centre line of rural roads. For urban roads where shielding from noise due to adjacent properties and the distorting effects of gaps between properties and reflection will contain the noise only, properties within 0-49 and 50-99 m should be listed.

The visual impact of transport works

There can be little doubt that major transport works, and particularly urban highways, cause a considerable change in the visual scene. In some circumstances and to some people this causes a considerable loss of amenity.

In all aspects of transport, economic considerations are becoming increasingly important. Thus the cost of installation and operation of any transport scheme has to be weighed against alternatives, such as schools or welfare facilities, to which the resources could be allocated. In such a situation the extra cost of avoiding impact has to be balanced against the value of the amenity which is lost.

While the cost of improving the visual image of transport works or reducing impact by alternative routes or by tunnelling can be readily calculated, the reduction in impact which results has been difficult to estimate.

In the context of trunk roads the Standing Advisory Committee on Trunk Road Assessment[10] distinguished two types of visual impact. When a road structure obstructs the view then this impact was referred to as visual obstruction. The more subjective effect of the road on the landscape was referred to as visual intrusion.

TABLE 10.2   Calculation of intensity and duration-weighted number of people subjected to noise impact

	Scale score S	Proportion with scale score S	No. of people	Intensity weighting	Duration weighting	Weighted number of people
Residential development	1	0.19	114	+1	1.0	+114
	2	0.18	108	0	1.0	0
	3	0.16	96	−1	1.0	−96
	4	0.17	102	−2	1.0	−204
	5	0.09	54	−3	1.0	−162
	6	0.09	54	−4	1.0	−216
	7	0.12	72	−5	1.0	−360
School	1	0.15	45	+1	0.5	+22
	2	0.16	48	0	0.5	0
	3	0.13	39	−1	0.5	−20
	4	0.18	54	−2	0.5	−54
	5	0.12	36	−3	0.5	−54
	6	0.11	33	−4	0.5	−66
	7	0.15	45	−5	0.5	−112
Church	1	0.08	16	+1	0.2	+4
	2	0.08	16	0	0.2	0
	3	0.09	18	−1	0.2	−4
	4	0.14	28	−2	0.2	−11
	5	0.18	36	−3	0.2	−22
	6	0.17	34	−4	0.2	−27
	7	0.26	52	−5	0.2	−52
						−1320

Some initial research into the measurement of visual obstruction has been carried out by Hopkinson[11] who has proposed a tentative method by which an assessment of visual obstruction can be made.

It is possible to measure some forms of obstruction without too much difficulty. If, for example, a large bridge structure is erected immediately in front of a dwelling house, then the loss of daylight is a reasonable measure of the obstruction. In a great many cases, however, there is no loss of daylight, and it is simply the view which is changed. Where a new highway or public transport interchange replaces derelict urban property, many residents may consider the intrusion an improvement in amenity. In a suburban area the same construction, requiring the demolition of residential properties, would usually be considered to be an intrusion with a considerable loss of amenity.

When considered in this way it would seem that the problem defies rational solution but the problem of noise impact presents similar difficulties. The effect of intruding noises varies according to the frequency and level of the intrusion and in this situation measurement scales have been calibrated against human response as obtained from social surveys.

An attempt to assess the visual consequences of transport works was first made by Hopkinson, who used the principle that the extent of visual obstruction can be quantified by the amount of the field of view taken up by the obstruction. A unit by which this field of view may be measured is the solid angle expressed in terms of the steradian, the angle subtended at the centre of a sphere of unit radius by a unit area on its surface. In practice the millisteradian (msr) is the unit used.

An additional factor which must be considered in assessing the obstruction of an element in the field of view is the position of the object in the visual field. This is measured by the position factor: an object subtending a given angle at the centre of the field of view is considered to be more obstructive than an object subtending the same angle near the periphery of the visual field. The central zone up to 6° from the centre of the field of view has the greatest significance and is given a weighting of 100; between 6° and 20° the position factor is 30; from 20° to 50°, the outer limit of which covers most of the binocular field, the position factor is 10. From 50° to 90° is that part of the visual field seen by only one or other of the eyes, and has a position factor of 1. The solid angular subtense of an object may then be multiplied by the position factor to give a weighted solid angle.

For the simplified case of a straight elevated highway or railway embankment, or similar structure, with a uniform height H above the surrounding ground, it can be shown that the solid angle subtended by the transport facility is

$$S_\theta = \frac{H}{d_p} \ (\cos\theta_1 - \cos\theta_2)$$

where d_p is the perpendicular distance in plan from the point at which the visual obstruction is being measured to the line of the transport facility; θ_1 and θ_2 are the angles subtended between the rays to the limits of vision of the works and a datum line parallel to the road. This is only approximately correct; where the viewpoint is close to the intrusion, so that the vertical angle subtended at the viewpoint is greater than approximately 30°, a spherical correction is required.

The height H of the works depends largely on the height of the structure when it is elevated, but in the case of depressed facilities the height may be the barrier wall or, for open cuttings, the perceived height of the open excavation.

Determination of the solid angle may be made by calculation after measurement from plans and sections of the proposed works, by the use of special full-field cameras or by the use of specially devised protractors which make the employment of non-specialist staff possible.

Calculation from measurements of H, d_p, $\cos\theta_1$ and $\cos\theta_2$ made on plans and sections is tedious. Where a preliminary impact analysis is being carried out it may be sufficient to calculate the solid angle subtended by the central 40° of the field of view and ignore the effect of the position factor.

The impact of air pollution

Motor vehicles propelled by engines using petrol, petroleum gas or diesel as a fuel emit a wide range of gaseous and particulate materials some of which have a potential to be harmful to human beings, the amount of the pollution depending on the condition and type of the engine and on operating conditions.

There is increasing public disquiet regarding the effects of motor vehicle air pollution and this has resulted in strict control of exhaust emissions in the United States. From 1971 progressively more stringent regulations have been introduced in Europe governing the emissions of passenger cars. These regulations cover carbon monoxide, hydrocarbons and oxide of nitrogen levels emitted under standard test conditions. Over the same period the total amount of lead emitted has been reduced by controlling the proportion added to petrol as a means of increasing engine efficiency.

Public reaction to air pollution likely to be caused by proposed highway schemes is often expressed by residents of communities adjacent to the line of new road schemes. This is a growing problem as many highway proposals are, in the future, likely to be in urban areas.

This concern has been noted by the Advisory Committee on Trunk Road Assessment[12] which recommended that "where air pollution is likely to be a problem a special air quality report should be prepared: otherwise it should be excluded from the assessment." This view was endorsed by the Standing Advisory Committee on Trunk Road Assessment, which commented that the impact of air pollution should be assessed where it is a particular problem.

For these reasons the Department of Transport in their Manual of Environmental Appraisal[13] recommend that where appropriate an air quality report should be included in the environmental assessment of alternative proposals. The subject of air pollution due to road traffic is considered in detail in Section 26.

1. Department of Transport, *The COBA Manual*, London (1981)
2. J. C. Tanner, Forecasts of vehicle ownership in Great Britain, *Roads and Road Construction*, **42** (1965), 515, pp 341-7; **43** (1965), 516, pp 371-6
3. J. C. Tanner, Forecasts of vehicles and traffic in Great Britain, 1974 revision, *Transport and Road Research Laboratory Report* LR 650, Crowthorne (1974)
4. J. C. Tanner, Car ownership trends and forecasts, *Transport and Road Research Laboratory Report* LR 799, Crowthorne (1977)
5. Departments of the Environment and Transport, A disaggregate model of household car ownership, *Research Report 20*
6. Department of Transport, *National Road Traffic Forecasts* (1984)
7. L. B. Leopold, F. E. Clarke, B. B. Hanshaw, and J. R. Balsey, A procedure for estimating environmental impact, *U.S. Dept. of Interior Geological Survey Circular No. 645* (1971)
8. Department of Transport, Frameworks for trunk road appraisal, *Departmental Standard TD/12/83*, London (1983)
9 Department of the Environment, *The Environmental Evaluation of Transport Plans*, HMSO, London (1976)

10. Department of Transport, *Report of the advisory committee on trunk road assessment. Chairman: Sir G. Leitch*, HMSO, London (1977)
11. R. G. Hopkinson, The evaluation of visual intrusion in transport situations, *Traff. Engn. Control*, 387-91, 395
12. *Report of the Advisory Committee on Trunk Road Assessment. Chairman: Sir G. Leitch*, HMSO, London (1977)
13. Department of Transport, *Manual of Environmental Appraisal*, London (1983)

PART 2
ANALYSIS AND DESIGN
FOR HIGHWAY TRAFFIC

11

The capacity of highways between intersections

The capacity of a highway may be described as its ability to accommodate traffic, but the term has been interpreted in many ways by different authorities. Capacity has been defined as the flow which produces a minimum acceptable journey speed and also as the maximum traffic volume for comfortable free-flow conditions. Both these are practical capacities while the Highway Capacity Manual[1] defines capacity as the maximum hourly rate at which persons or vehicles can reasonably be expected to traverse a point or uniform section of a lane or roadway during a given time period under prevailing roadway, traffic and control conditions. The time period used in most capacity analysis is 15 minutes which is considered to be the shortest interval during which stable flow exists.

Highway capacity itself is limited by:

1. The physical features of the highway, which do not change unless the geometric design of the highway changes.
2. The traffic conditions, which are determined by the composition of the traffic.
3. The ambient conditions which include visibility, road surface conditions, temperature and wind.

A term used in the Highway Capacity Manual to classify the varying conditions of traffic flow that take place on a highway is 'level of service'. The various levels of service range from the highest level, which is found at a flow where drivers are able to travel at their desired speed with freedom to manoeuvre, to the lowest level of service, which is obtained during congested stop-start conditions.

The Highway Capacity Manual approach

The level of service afforded by a highway to the driver results in flows that may be represented at the highest level by the negative exponential headway distribution when cumulative headways are being considered, by the double exponential distribution as the degree of congestion increases, and by the regular distribution in 'nose-to-tail' flow conditions.

To define the term level of service more closely, the Highway Capacity Manual gives six levels of service and defines six corresponding volumes for a number of highway types. These volumes are referred to as maximum service flow rates and may be defined as the maximum hourly rate at which persons or vehicles can reasonably be expected to traverse a point or uniform section of a lane or roadway during a given time period under prevailing roadway, traffic and control conditions while monitoring a designated level of services.

Levels of service are based on one or more operational parameters which describe operating quality for a particular type of facility. These parameters are referred to as measures of effectiveness and differ according to the type of facility, as detailed in table 11.1.

TABLE 11.1   Measures of effectiveness used for determination of level of service[1]

Facility	Measure of effectiveness
Freeways	
basic freeway segments	Density (pc/mi/ln)
weaving areas	Average travel speed (mph)
ramp junctions	Flow rates (pcph)
Multi-lane highways	Density (pc/mi/ln)
Two-lane highways	Percent time delay (%)
	Average travel speed (mph)
Signalised intersections	Average delay (sec/veh)
Unsignalised intersections	Reserve capacity (pcph)

Freeway operating characteristics include a wide range of rates of flow over which speed is relatively constant. For this reason, speed alone is not an adequate measure of performance for the definition of level of service. Drivers however are sensitive to the freedom to manoeuvre and proximity to other drivers, traffic flow characteristics which are related to the density of the traffic flow. Importantly, rate of flow is related to density throughout the range of traffic flow. For this reason density is now the parameter used to define level of service for basic freeway segments.

Descriptions of operating conditions for the six levels of service are as follows.

Level of service A is primarily free-flow operation with average travel speeds near 60 mph on 70 mph freeways and vehicles have almost complete freedom to manoeuvre. Minor traffic incidents and breakdowns are easily absorbed and traffic quickly returns to level of service A. This level of flow gives drivers a high level of physical and psychological comfort. Average spacing between vehicles gives a maximum density of 12 passenger cars per mile per lane.

Level of service B also represents reasonable free-flow conditions and speeds of over 57 mph are maintained on 70 mph freeways, and vehicles have only a slightly restricted freedom to manoeuvre. Minor traffic incidents and breakdowns are still easily absorbed although local deterioration in flow conditions would be more marked than in level of service A. The psychological and physical level of service is still good. Maximum density is 20 passenger cars per mile per lane.

Level of service C gives stable operation but flows approach the range in which a small increase in flow will cause a marked reduction in service; average speeds over

54 mph are maintained. Freedom to manoeuvre is considerably restricted and lane changes require care. Incidents may still be absorbed but local deterioration in service will be substantial with queues forming at major incidents. Drivers experience a major increase in tension because of the increased density of traffic flow which has a maximum value of 30 passenger cars per mile per lane.

Level of service D borders on unstable flow and small increases in flow result in substantial falls in the level of service; average speeds are in the region of 4-6 mph. Freedom to manoeuvre is severely limited with drivers experiencing drastically reduced physical and psychological levels. Maximum density is 42 passenger cars per mile per lane.

Level of service E — the boundary between level of service D to level of service E describes capacity operation; traffic flow at this level is very unstable, vehicles are spaced at approximately uniform headways, any disruption due to a vehicle entering the traffic stream forms a disruptive wave which moves upstream, and average speeds are in the region of 30 mph. Any incident produces extensive queueing. Physical and psychological conditions afforded to drivers are extremely poor. Maximum density has a maximum value of 67 passenger cars per mile per lane. Level of service F describes breakdown or forced flow, conditions behind points on the highway where demand exceeds capacity.

The Highway Capacity Manual gives values of maximum service for freeways and these are given in table 11.2. These rates represent the maximum service flow rates for traffic flow in ideal conditions for 70, 60 and 50 mph design speed freeways.

The values for speed given in table 11.2 reflect the influence of a 55 mph speed limit and, unless enforcement is stringent, average travel speeds are expected to be slightly higher than the speed limit. The maximum service flow rates are given in units of passenger cars per hour per 12 ft lane.

The service flow rate for a particular roadway element is then obtained by muliplying by the number of lanes in one direction of the freeway, a factor to adjust for the effects of restricted lane width and lateral clearance, a factor to adjust for the effect of heavy vehicles, and a factor to adjust for the effect of driver population.

The factor which accounts for restricted lane width and/or lateral clearance takes account of lane widths narrower than 12 ft and/or for objects closer to the edge of the travel lanes than 6 ft either at the roadside or in the central reservation. Abridged details are given in table 11.3 and in the use of this table it is recommended that judgement be made regarding the effect of crash barriers which do not produce an effect on traffic flow.

Adjustment for vehicle type in the Highway Capacity Manual is made by determining the passenger-car equivalent for each truck, bus or recreational vehicle. The Highway Capacity Manual gives values of the passenger car equivalents of trucks and recreational vehicles depending upon the severity and length of the gradient, the percentage of trucks and recreational vehicles in the traffic stream and the highway type. For freeways, trucks are divided into three classes depending upon the power-to-weight ratio. When capacity analysis is being carried out for considerable lengths of freeway, then grade conditions can be generally classified as level, rolling or mountainous terrain. The definition of rolling terrain is where any combination of horizontal or vertical alignment causes heavy vehicles to reduce their speed substantially below those of passenger cars, but does not cause heavy vehicles to travel at crawl speeds for any significant length of time. Mountainous terrain, on the other hand, causes heavy vehicles to travel at crawl speeds for significant distances or at

TABLE 11.2   Maximum service flow rates[1]

Design speed (mph)	Level of service	Density (pc/mi/ln)	Speed (mph)	Maximum service flow rate
	A	$\leqslant 12$	$\geqslant 60$	700
	B	$\leqslant 20$	$\geqslant 57$	1100
	C	$\leqslant 30$	$\geqslant 54$	1550
70	D	$\leqslant 42$	$\geqslant 46$	1850
	E	< 67	$\geqslant 30$	2000
	F	> 67	< 30	unstable
	B	$\leqslant 20$	$\geqslant 50$	1000
	C	$\leqslant 30$	$\geqslant 47$	1400
60	D	$\leqslant 42$	$\geqslant 42$	1700
	E	< 67	$\geqslant 30$	2000
	F	> 67	< 30	unstable
	C	$\leqslant 30$	$\geqslant 43$	1300
	D	$\leqslant 42$	$\geqslant 40$	1600
50	E	< 67	$\geqslant 28$	1900
	F	> 67	< 28	unstable

frequent intervals. Using this broad classification, the passenger car equivalents given in table 11.4 may be used.

The adjustment for driver population takes account of the observed differences in characteristics between regular weekday and commuter drivers and weekend drivers especially in recreational areas. It is recommended that a factor of between 0.75 and 0.90 be used to account for the different characteristics of drivers who are not regular weekday or commuter drivers.

The United Kingdom approach

In Great Britain, carriageway provision both in type and width depends upon an assessment of traffic flow forecasts, the composition of the traffic, variations in traffic flow referred to as 'peaking', the costs of delays during traffic incidents and maintenance, and environmental effects[2].

To determine appropriate carriageway widths for rural roads, the Department of Transport[3] have made extensive assessments of the economic benefits of providing

TABLE 11.3 Adjustment factor for restricted lane width and lateral clearance[1]

Distance from travelled carriageway (ft)	Obstruction on one side of road			Obstruction on both sides of road		
	Lane width (ft)					
	12	11	10	12	11	10
	4-lane freeway (2 lanes each direction)					
⩾ 6	1.00	0.97	0.91	1.00	0.97	0.91
4	0.99	0.96	0.90	0.98	0.95	0.89
2	0.97	0.94	0.88	0.94	0.91	0.86
0	0.90	0.87	0.82	0.81	0.79	0.74
	6 or 8-lane freeways (3 or 4 lanes each direction)					
⩾ 6	1.00	0.96	0.89	1.00	0.96	0.89
4	0.99	0.95	0.88	0.98	0.94	0.87
2	0.97	0.93	0.87	0.96	0.92	0.85
0	0.94	0.91	0.85	0.91	0.87	0.81

TABLE 11.4 Passenger car equivalents for extended freeway segments[1]

Passenger-car equivalent	Terrain		
	level	rolling	mountainous
For trucks	1.7	4.0	8.0
For buses	1.5	3.0	5.0
For recreational vehicles	1.6	3.0	4.0

different carriageway widths on new rural roads. Time, vehicle operating and accident costs were assessed using cost–benefit analysis. In this analysis, estimates of total maintenance costs were made by assuming typical work schemes and associated delays; costs of traffic incidents in terms of delay to other vehicles were also included.

Economic assessment indicated the lower bound of a flow range, the lowest at which a given carriageway width was likely to be preferred to a lesser width. The upper bound of a flow range is however determined not only by economic assessment but also by operational assessment. This form of assessment indicates the maximum traffic flow which a given width can accommodate under some stated conditions.

Two factors were considered when economic assessments were modified by operational assessment. These were: the duration of those periods within the traffic peaks when the flow was greater than the road 'capacity' and the diversion of flows during maintenance.

In the assessment process the projected growth of traffic in the future can result in traffic flows being reached which exceed the maximum flow levels which have actually been observed on the highway. In the economic assessment it is assumed that when this takes place the costs which are incurred are as a result of traffic diverting to other routes or being suppressed. This is considered satisfactory for short-lived peak flows but is unrealistic for longer periods of time. For this reason, in the operational assessment 'over-capacity' was confined to peak flow periods in the thirtieth year of assessment. It was found that this limitation reduced the upper bound of dual carriageway and motorway flow ranges.

The second factor considered was the limitation on flows caused by a maximum possible flow which could be diverted during maintenance works. For single carriageway roads the maximum divertable flow was assumed to be approximately 2000 vehicles per day, and 10 000 vehicles per day for dual carriageways and motorways.

As a result of these assessments, the flow levels for rural road assessment have been published by the Department of Transport[3] and are given in table 11.5.

The unit which is used to measure traffic flow in this table is the 24-hour Annual Average Daily Traffic which is the total annual traffic on a road divided by 365. The traffic flow is expressed in vehicles in table 11.5; an allowance for the composition of the traffic flow is made in the economic assessment of the adoption of differing

TABLE 11.5   Flow levels for the assessment of rural roads[3]

Road class	AADT 15th year after opening	Access treatment
Normal single 7.3 m carriageway	up to 13 000	Restriction of access, concentration of turning movements, clearway at top of range
Wide single 10 m carriageway	10 000 to 18 000	Restriction of access, concentration of turning movements, clearway at top of range
Dual 2-lane all-purpose carriageways	11 000 to 30 000	Restriction of access, concentration of turning movements, clearway at top of range
	30 000 to 46 000	Severe restriction of access, left turns only, clearway
Dual 3-lane all-purpose carriageways	40 000 and above	Severe restriction of access, left turns only, clearway
Dual 2-lane motorway	28 000 to 54 000	Motorway regulations
Dual 3-lane motorway	50 000 to 79 000	Motorway regulations
Dual 4-lane motorway	77 000 and above	Motorway regulations

carriageway types. Essential to this is the effect of vehicle type on the speed/flow relationship; this will be discussed separately. As indicated in table 11.5, traffic growth is taken into account by designing for the traffic flows which are expected in the 15th year after opening of a road.

The flow ranges given in table 11.5 are not intended to be used inflexibly; they are a starting point for the assessment of a road proposal and merely meant to be a guide as to which road layouts are likely to be operationally and economically acceptable in normal circumstances. In the United Kingdom, the Department of Transport do not stipulate a minimum level of service in road design; decisions on road capacity must reflect economic and environmental considerations. Where construction and environmental costs are high, then the incremental cost between two alternative road widths may result in a lower standard of provision than might otherwise have been made.

Alternatives for carriageway provision should be tested for assumptions of high and low traffic growth by considering the effects on travellers in terms of using

TABLE 11.6   Design flows of two way urban roads[3]

Road type	Carriageway		Peak hourly flow (veh/hour)
	type	width (m)	
Urban	2-lane dual	7.3 dual	3600
motorway	3-lane dual	11 dual	5700
All purpose, no frontage access,	2-lane carriageway	7.3	2000*
no standing veh., negligible cross	2-lane carriageway	10	3000*
traffic	4-lane undivided	12.3	2550
	4-lane undivided	13.5	2800
	4-lane undivided	14.6	3050
	2-lane dual	6.75 dual	2950†
	2-lane dual	7.3 dual	3200†
	2-lane dual	11 dual	4800†
All purpose,	2-lane carriageway	6.1	1100*
frontage development,	2-lane carriageway	6.75	1400*
side roads,	2-lane carriageway	7.3	1700*
pedestrian	2-lane carriageway	9	2200*
crossings,			
bus stops,	2-lane carriageway	10	2500*
waiting			
restrictions	4-lane undivided	12.3	1700
throughout day,	4-lane undivided	13.5	1900
loading restrictions	4-lane undivided	14.6	2100
at peak hours	6-lane undivided	18	2700

*Peak hourly flow, both directions of flow.
†60/40 directional split assumed.

travel cost savings, accident savings, user costs during maintenance, reconstruction, accidents and breakdowns, and driving conditions. Occupiers of property within 300 m which are subject to increased visual impact, the number of properties requiring demolition, and agricultural land acquisition should be included in the consideration. Transport development policy decisions which affect carriageway width are taken into account and the financial implications considered.

Design flows for urban roads are based on peak hour flows and specifically apply to roads which function as traffic links and are independent of the capacities of junctions. There may be circumstances where traffic management policy requires lower values of design flows for environmental and safety reasons. In determining the peak hour flow, this is defined as the highest flow for any specific hour of the week averaged over any consecutive 13 weeks during the busiest period of the year. In the United Kingdom the busiest 3-month period is likely to be June to August but considerable differences occur, particularly in urban areas; the weekday peak hour is normally 5–6 p.m. on Friday, but once again local variations are possible. Practically, measurements over 13 weeks are not essential and a period of 5 to 7 weeks is normally considered adequate.

Design flows for urban roads given by the Department of Transport[3] are detailed in tables 11.6 and 11.7. In adopting the values, the following points should be kept in mind except when dual carriageway links are being considered. Firstly, in most cases the design flows of existing roads will be dependent on the capacity of terminal junctions. Secondly, although the road types given cover a range of conditions which are difficult to define precisely, extreme conditions can be met where the design flows given cannot be attained. Thirdly, the design flows are only appropriate when the road is used solely as a traffic link.

The design flows given in tables 11.6 and 11.7 allow for a heavy vehicle content in the flow of 15 per cent, and no adjustment is necessary for lower heavy vehicle contents. Where the heavy vehicle content exceeds 15 per cent, then a correction to the design flow as given in table 11.8 should be made.

TABLE 11.7　Design flows for one way urban roads[3]

Road type	Carriageway width (m)	Design flow, one direction of flow (veh/hour)
All purpose road,	6.75	2950
no frontage access,	7.3	3200
no standing veh.,	11	4800
negligible cross traffic		
All purpose road,	6.1	1800
no frontage development,		
side roads,	6.75	2000
pedestrian crossings,	7.3	2200
bus stops,	9	2850
waiting restrictions		
throughout day,	10	3250
loading restrictions at peak hours	11	3550

TABLE 11.8 Corrections to tables 11.6 and 11.7 for heavy vehicle content[3]

Road type	Heavy vehicle content (%)	Total reduction in flow (veh/hour)
Motorway and dual carriageway all purpose roads	15–20	100
Motorway and dual carriageway all purpose roads	20–25	150
10 m wide and above single carriageway roads	15–20	150
10 m wide and above single carriageway roads	20–25	225
Below 10 m wide single carriageway roads	15–20	100
Below 10 m wide single carriageway roads	20–25	150

The effect of journey time on design flows

In the selection of a design flow range in the United Kingdom, an economic assessment is made of the alternative carriageway widths which are suggested in table 11.5. Essential to such an assessment is the estimation of travel speed for each carriageway width for varying traffic flow, vehicle composition and highway geometry. The assessment is made using the Department of Transport COBA[4] program which incorporates travel speed relationships. Different speed prediction relationships are used according to road types: rural single carriageways, rural all-purpose dual carriageways, motorways, suburban, and urban roads.

On rural roads the speed is predicted for each link and delays at intersections estimated separately; in urban areas the road system, including intersections, has to be considered as a network. Separate relationships are given for light and heavy vehicles. Light vehicles are defined as cars and delivery vans, and vans of larger carrying capacity but excluding vehicles with twin rear types. Heavy vehicles are defined as all other goods vehicles, buses and coaches.

For rural single carriageway roads the factors which influence travel speed are detailed below.

(a) Bendiness, the total change of direction per unit distance (deg/km).
(b) Hilliness, total rise and fall per unit distance (m/km).
(c) Net gradient, net rise per unit distance, used only for one-way links (m/km).
(d) Total flow, all vehicles per standard lane (veh/hour/3.65 m lane).
(e) Average carriageway width (nominal 7.3 m or 8.5 m and above).
(f) Average verge width, both sides, including any metre strips (m).
(g) Total number, both sides, of laybys, side roads and accesses excluding house and field entrances per km (no./km).
(h) Average sight distance (harmonic mean) (m).

For rural all-purpose dual carriageways and motorways the factors which influence travel speed are:

(a) Bendiness, the total change of direction per unit distance (deg/km).
(b) Sum of rises per unit distance (m/km).
(c) Sum of falls per unit distance (m/km).
(d) Total flow, all vehicles per standard lane (veh/hour/3.65 m lane).

For central area urban roads, average off-peak speed of all vehicles varies with the frequency of major intersections per km. For non-central area urban roads, average off-peak speeds of all vehicles vary with the degree of development which is defined as the proportion of the non-central road network that has frontage development. Shops, offices and industry are counted as 100 per cent and residential development as 50 per cent. The figure used should be the weighted average for all links in the non-central area.

Suburban roads, defined as major suburban routes in towns and cities where the speed limit is generally 40 mph. For these roads the average journey speed of all vehicles includes delays at junctions and varies with the following variables:

(a) Frequency of major intersections (no./km).
(b) Number of minor intersections and private drives per km.
(c) Percentage of development.
(d) Percentage of heavy vehicles.
(e) Total flow, all vehicles per standard lane (veh/hour/3.65 m lane).

References

1. Transportation Research Board, Highway Capacity Manual, *Special Report* 209, Washington DC (1985)
2. Department of Transport, Roads and Local Transport Directorate, Choice between options for trunk road schemes, *Departmental Advice Note* TA 30/82, London (1982)
3. Department of Transport, Highways and Traffic. Traffic flows and carriageway width assessment for rural roads, *Departmental Advice Note* TD 20/85. London (1985)
4. Department of Transport, *The COBA program*, London (1981)

Problems

Are the following statements true or false.

(a) The selection of carriageway width for a given design flow should in the United Kingdom be determined by a consideration of economic factors.
(b) The derivation of maximum service flow rates in the Highway Capacity Manual is based upon average travel speeds.
(c) In the United Kingdom a preliminary starting point for the assessment of carriageway width in rural areas is the maximum observed peak hour flow.
(d) The Highway Capacity Manual defines level of service A as representing congested nose-to-tail driving conditions.

(e) Design flow ranges recommended for used in the United Kingdom were derived from a consideration of minimum travel cost.

(f) Crash barriers adjacent to the travelled carriageway inhibit the behaviour of drivers and reduce design flow levels.

(g) The effect of traffic composition is taken into account when calculating design flow levels by the use of passenger car equivalents.

Solutions

(a) In United Kingdom practice the recommended design flow ranges will normally allow a highway designer a choice of types and widths. The choice of option is made by a consideration of user travel costs, accident costs, user costs during maintenance, reconstruction, after accidents and vehicle breakdowns, and also driving conditions on the highway. In addition environmental effects, property demolition and land acquisition must be considered. All these effects must be evaluated for both high and low traffic growth forecasts.

(b) In United States practice the Highway Capacity Manual relates design flows to the expected level of service which is to be provided by the highway to the road user. For freeways these levels of service are related to traffic density, a traffic flow parameter which is related to traffic conditions as observed by drivers.

(c) In United Kingdom practice the starting point for the assessment of alternative carriageway widths is based on design flow ranges which are expressed by the Annual Average Daily Traffic, that is, the total flow of all vehicles in a year divided by 365. The fact that flows vary considerably both during the day and the year is taken into account when user costs are estimated by the Department of Transport COBA program. This program uses flow groupings, varying for differing road categories, which take into account the variations of flow throughout the year and the differing proportions of vehicle types as the flow varies.

(d) The Highway Capacity Manual level of service A represents free-flow, not congested, traffic conditions. Vehicles have almost complete freedom to manoeuvre, giving drivers a high level of physical and psychological comfort.

(e) Design flow ranges used for determining carriageway width in the United Kingdom are determined for the lower end of the range from economic considerations using cost–benefit analysis whilst at the upper end economic considerations are modified by operational considerations, that is, traffic delays during maintenance, the effects of traffic incidents and accidents, and the maximum flow of traffic which can be diverted from a highway during obstructions to flow.

(f) Whilst the Highway Capacity Manual gives corrections for design flows that arise because of continuous obstructions adjacent to the carriageway, the presence of crash barriers does not appear to inhibit driver behaviour and it is recommended that in this case judgement be used in applying a correction.

(g) The effect of different vehicle types on design flows is taken into account in the Highway Capacity Manual by assigning an equivalent passenger car value to each vehicle type. In United Kingdom practice the cost–benefit analysis which assists in carriageway width choice divides vehicles into cars, public service vehicles, light goods vehicles and two classes of other goods vehicles. Design flows are expressed in vehicles and are not converted into passenger car units.

12

Headway distributions in highway traffic flow

The concept of level of service in highway traffic flow illustrates the differences in the characteristics of the flow which may be examined by a study of the headways between vehicles. Time headways are the time intervals between the passage of successive vehicles past a point on the highway. Because the inverse of the mean time headway is the rate of flow, headways have been described as the fundamental building blocks of traffic flow. When the traffic flow reaches its maximum value then the time headway reaches its minimum value.

If time headways are observed during any period of time the individual values of time headway vary greatly. The extent of these variations depends largely on the highway and the traffic conditions.

On a lightly trafficked rural motorway, where vehicles can overtake at will, a range of headways will be observed from zero values between overtaking vehicles to the longer headways between widely spaced vehicles.

When flow conditions are observed on more heavily trafficked highways there are fewer opportunities to overtake and fewer more widely spaced vehicles. When overtaking opportunities do not exist there is an absence of very small headways and, under very heavily trafficked conditions, all vehicles are travelling at uniform headways as they follow each other along the carriageway.

There are two general approaches to the method of measurement of headways. They may be measured by a device that registers the successive arrivals of vehicles at a fixed point. Alternatively headways may be recorded by aerial photography which records at one instant of time the distribution of headways between successive vehicles.

By the first method it is the time headway distribution that is obtained and in the second method it is the space headway distribution. Because of the ease of observation it is the time headway distribution that has been extensively researched and reported.

When the arrival of vehicles at a particular point on the highway is described, the distribution may either describe the number of vehicles arriving in a time interval or the time interval between the arrival of successive vehicles. The first type is the counting distribution and the second type is the gap distribution.

100

As great variability in all types of headway may be recorded, it has been described as the most noticeable characteristic of vehicular traffic. For this reason attempts to understand headways have employed statistical methods and probability theory to find theoretical distributions to represent the observed headway distributions.

Haight, Whisler and Mosher[1] studied the relationship between the counting and the gap distribution but usually it is the gap distribution that is studied. It requires a shorter period of observation to collect data for the investigation of gap distributions than for an investigation into counting distributions. In the study of intersection capacity, gaps in the major road flow are used by minor road vehicles to enter the major road and, once again, it is the gap distribution that is of importance.

One of the earliest headway distributions proposed for vehicular traffic flow was proposed by Kinzer[2] and Adams[3] who suggested that the negative exponential distribution would be a good fit to the cumulative gap distribution. Adams illustrated the validity of the negative exponential distributions by observations of traffic flow in London. When this distribution represents the cumulative headway distribution then arrivals occur at random and the counting distribution may be represented by the Poisson distribution. This type of flow may be found where there are ample opportunities for overtaking, at low volume/capacity ratios.

The negative exponential headway distribution

If the traffic flow is assumed to be random then the probability of exactly n vehicles arriving at a given point on the highway in any t second interval is obtained from the Poisson distribution which states

$$\text{probability } (n \text{ vehicles}) = (qt)^n \exp(-qt)/n! \tag{12.1}$$

where q is the mean rate of arrival per unit time.

Often this distribution is referred to as the counting distribution, because it refers to the number of vehicles arriving in a given time interval. It is the negative exponential distribution however which is most commonly used when describing headway distributions.

The negative exponential distribution can be obtained from the Poisson distribution if there are no vehicle arrivals in an interval t. In this case there must be a headway greater than or equal to t.

$$\text{probability } (\text{headway} \geqslant t) = \exp(-qt) \tag{12.2}$$

The mean rate of arrival q is also the reciprocal of the mean headway which can be computed from the observed headways.

It is possible to demonstrate the use of the Poisson and negative exponential distributions by taking observations of headways on a highway where traffic is flowing freely. This flow condition is likely to be found on two-way two-lane highways when the traffic volume in each direction does not exceed 400 veh/h or 800 veh/h on one-way two-lane carriageways and there are no traffic control devices within a distance of approximately 1 km upstream of the point of observation. The time interval between the passage of successive vehicles is noted using a stopwatch and the resulting headways are placed into classes with a class interval of 2 seconds. A smaller class interval is desirable but it is not normally justified unless more accurate means of measuring the headways are available. It will normally be necessary to continue observations for 30 minutes to obtain sufficient headways.

Observations, derived and theoretical values may be tabulated as shown in table 12.1.

TABLE 12.1

Row 1	Headway class
Row 2	Observed frequency of headways in class
Row 3	Observed frequency of headway $\geqslant$ lower class limit
Row 4	Row 3 expressed as a percentage
Row 5	Theoretical percentage of headways $\geqslant$ lower class limit (equation 12.2 multiplied by 100 where $t = 0, 2, 4 \ldots$)
Row 6	Theoretical frequency of headways $\geqslant$ lower class limit (row 5 multiplied by total number of headways observed)
Row 7	Theoretical frequency of headways in class (obtained by difference between successive values of row 6)

It should be noted that the following rows represent observed and theoretical values

row 2 and row 7

row 3 and row 6

row 4 and row 5

An example of observed and theoretical values is given in table 12.2. This table is incomplete as the greatest headway was in the class 64 to 65.9 seconds, but for the sake of brevity only a limited number of values are given.

TABLE 12.2

Row 1	0	2	4	6	8	10	12	14	16
Row 2	34	27	16	13	13	10	6	7	5
Row 3	168	134	107	91	78	65	55	49	42
Row 4	100	80	64	54	46	39	33	29	25
Row 5	100	84	70	58	49	41	34	29	24
Row 6	168	140	117	98	82	68	57	48	40
Row 7	28	23	19	16	14	11	9	8	7

Note that differences between successive values in row 3 give values in row 2; similarly differences in row 6 give values in row 7. This is because the number of headways greater than t_1 second minus the number of headways greater than t_2 second gives the number of headways in class $t_1 t_2$ second.

The closeness of fit of the negative exponential distribution to the observed values can be demonstrated graphically by drawing a histogram of the observed and theoretical number of headways in each class as in figure 12.1.

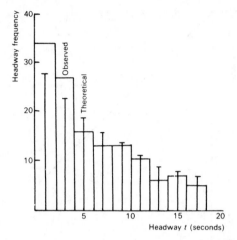

Figure 12.1   A comparison of observed and theoretical headway frequencies

However a statistical test of the closeness of fit of the theoretical distribution can be made by means of the chi-squared test, where chi-squared is the sum for each headway class of the following statistic

$$\frac{(\text{observed frequency} - \text{theoretical frequency})^2}{\text{theoretical frequency}}$$

that is

$$\frac{(\text{row 2} - \text{row 7})^2}{\text{row 7}}$$

Where the theoretical or observed frequency is less than 8 in any class, that class is combined with subsequent classes until the combined theoretical and observed frequencies are greater than 8.

Chi-squared is calculated in table 12.3.

TABLE 12.2   Calculation of closeness of fit

Row 2	Row 7	$(\text{Row 2}-\text{row 7})^2/\text{row 7}$
34	28	1.28
27	23	0.69
16	19	0.47
13	16	0.56
13	14	0.07
10	11	0.09
18*	24*	1.50 = 4.66

*Last three classes combined.

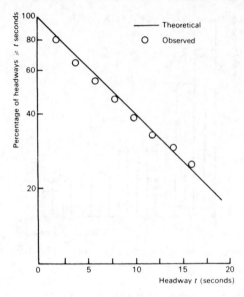

Figure 12.2   A comparison of observed and theoretical cumulative headway distributions

Examination of tables of chi-squared shows that at the 5 per cent level the value of chi-squared for 5 degrees of freedom is 11.07. The calculated value is 4.66 showing that there is no significant difference between the observed and theoretical headway distribution at the 5 per cent level.

Note that the degrees of freedom are equal to the number of combined classes, in this case 7, minus the number of parameters that were used from the observed to calculate the theoretical values. These two parameters are the total number of observed headways and the mean headway; the former was used in the calculation of row 6 and the latter in the calculation of row 5.

Another way in which the fit of the observed distribution to the theoretical distribution can be demonstrated is by plotting the theoretical and observed cumulative percentage frequency. These values are given in rows 4 and 5 respectively. If equation 12.2 is considered and natural logarithms are taken of both sides of the equation, then

$$\ln \left[\text{probability (headway} \geq t) \right] = -qt \tag{12.3}$$

When plotted on semi-log paper equation 12.3 is a straight line; the nearer the observed distribution is to a straight line the better is the fit. This is illustrated graphically in figure 12.2.

This experiment shows, for the traffic volume observed and for the class interval chosen, that the distribution of headways may be represented by the exponential distribution. It should be noted that one of the difficulties of the exponential distribution, the surplus of small headways less than 1 second, is hidden by the use of a class interval of 2 seconds.

Congested vehicular headway distributions

While free-flowing vehicular headway distributions may be approximated by the negative exponential distribution in the case of the cumulative headway distribution and by the Poisson distribution in the case of the counting distribution, it is obvious that the distribution of headways will depend on the traffic volume and also on the capacity of the highway. If drivers cannot maintain their desired speed by overtaking slower moving vehicles then free-flow conditions no longer exist and the highway is beginning to show signs of congestion.

Highway congestion may increase until finally all vehicles are travelling at the same speed and following each other at their minimum headway.

In this case vehicles are regularly distributed along the highway and the headway distribution is determinate so that,

$$\text{probability (headway} \geq t) = 0 \quad \text{when} \quad t \neq \bar{t}$$

$$= 1 \quad \text{when} \quad t = \bar{t}$$

where $\bar{t}$ is the mean headway between vehicles; the counting distribution is also given by

$$\text{probability } (n \text{ vehicles in time } t) = \frac{t - N\bar{t}}{\bar{t}}$$

where $n = N + 1$, and

$$= 1 - \frac{t - N\bar{t}}{\bar{t}}$$

where $n = N$.

N is the number of headways of time headway $\bar{t}$ contained in time interval t.

It is not however so much the two extremes of regularity or randomness which is of interest in the study of traffic flow as the distribution of these headways in the intermediate conditions between these extremes. The study of traffic flow requires a knowledge of the distribution of headways in a wide variety of traffic conditions if realism is to be achieved.

A difficulty of the use of the negative exponential distribution even under free-flow conditions is that the probability of observing a headway increases as the size of the headway decreases. As vehicles have a finite length and a minimum following headway this presents a problem when only a limited number of overtakings are observed. For this reason traffic flow has been described by the use of the displaced exponential distribution. This is a suitable distribution for the description of flow in a single lane of a multilane highway.

Where a small number of low-value headways are observed, such as is the case when a limited amount of overtaking is possible, the Pearson type III or the Erlang distribution may be used to represent the headway frequency distribution.

The Pearson type III distribution was developed by Karl Pearson to deal with a wide variety of statistical data. A random variable such as a headway (t) is said to be distributed as the type III distribution if its probability density function is given by

$$f(t) = \frac{b^a}{\Gamma(a)} \, t^{a-1} \, e^{-bt} \quad 0 < t < \infty \tag{12.4}$$

where $\Gamma(a)$ is known as the gamma function and is defined by

$$\Gamma(a) = \int_0^\infty z^{a-1} e^{-z} dz$$

Where headways in a single lane are being observed then any headway less than the minimum following distance cannot be observed and the Pearson type III distribution has to be modified to give

$$f(t) = \frac{b^a}{\Gamma(a)} (t - c)^{a-1} e^{-b(t-c)} \quad c < t < \infty \tag{12.5}$$

For equation 12.4 the theoretical distribution may be obtained from

$$\bar{t} = \frac{a}{b} \quad \text{and} \quad s^2 = \frac{a}{b^2}$$

For equation 12.5 the theoretical distribution may be obtained from

$$\bar{t} = \frac{a}{b}, \quad s^2 = \frac{a}{b^2}$$

and c is the minimum observed headway.

The Erlang distribution is a simplified form of the Pearson type III distribution in which the parameter a is an integer. This may be written

$$f(t) = \frac{(qa)^a}{(a - 1)!} \times t^{a-1} e^{-aqt} \quad a = 1, 2, 3 \ldots \tag{12.6}$$

where q is the rate of flow, the reciprocal of the mean time headway.

The fitting of Pearson III and Erlang distributions to headway data

An example of the fitting of Pearson type III distributions is given below for headway data collected on the Leeds–Bradford highway at Thornbury, a one-way, two-lane highway. Headways were observed using a Marconi Speed Meter connected to a graphical recorder and the grouped data is shown in table 12.4.

The mean and the variance of the headway distribution is first calculated, giving

$$\bar{t} = 3.4 \text{ s} \quad \text{and} \quad s^2 = 7.9 \text{ s}$$

where $\bar{t}$ is the mean time headway and s^2 is the variance of the headways.

As headways are observed in the range 0–0.9 s the frequency curve may be assumed to pass through the origin, a reasonable assumption when it is considered that headways were observed on a one-way two-lane carriageway and frequent overtaking was possible. Using equation 12.5 to calculate the theoretical frequency distribution it is necessary to calculate the value of the parameters a and b.

From the relationship $\bar{t} = a/b$ and $s^2 = a/b^2$ then

$$a = 1.499 \quad \text{and} \quad b = 0.435$$

TABLE 12.4 Observed and fitted headway distributions

| Headway class (seconds) | Observed frequency | Theoretical frequency | |
		Equation 12.4 Pearson III	Equation 12.6 Erlang
0–0.9	57	75.3	51.0
1–1.9	99	83.6	85.5
2–2.9	69	69.7	80.0
3–3.9	58	53.3	62.4
4–4.9	44	39.4	45.1
5–5.9	25	27.9	30.8
6–6.9	14	19.7	20.4
7–7.9	13	13.9	13.2
8–8.9	11	9.4	8.4
9–9.9	6	6.6	5.2
10–10.9	3	4.5	3.2
11–11.9	3	2.9	2.0
12–12.9	3	2.1	1.2
13–13.9	0	1.2	0.7
14–14.9	1	0.8	0.4
15–15.9	4	0.4	0.2

The value of $\Gamma(a)$ was then calculated using a standard gamma function program, giving a value of 0.923. The value of the theoretical frequency may then be calculated using equation 12.5 by substituting for t the successive values of the mid-class marks.

Where a desk calculator or tables of the gamma function are not available the Erlang distribution, equation 12.6, may be used as a simpler approximation. The value of the parameter a in this distribution reflects the distribution of headways for a range of traffic flow conditions. When $a = 1$ the distribution becomes negative exponential and when a is infinite complete uniformity of headways results, the value of a thus reflecting flow conditions between free flowing and congested states.

For this example the value of a is chosen as 2 and using equation 6 the theoretical frequency distribution is evaluated and given in table 12.4, together with the observed headway frequencies.

A travelling queue headway distribution model

Miller[4] has described a traffic flow model whereby randomly placed vehicles are moved backwards in time where necessary in order to maintain a constant minimum headway. He derived a bunch length distribution which he verified by observation on a straight three lane section of the A4 between London Airport and Slough. In this case the queue length distribution is given by

$$P(n) = n^{n-1}r^{n-1} \exp(-rn)/n!$$

where the parameter r is given by bk, where k is the concentration of vehicles in the traffic stream and b is the mean distance headway of bunched vehicles.

It can be shown that

$$r = Bsq$$

where B is the mean time headway of bunched vehicles,

> s is the ratio of the mean speed of bunched vehicles to the mean speed of all vehicles,
> q is the flow.

Queue length distributions were computed by the author and the theoretical and observed queue lengths compared for flow on the Bradford to Wakefield Road when the traffic flow was 523 veh/h. Good agreement between observed and theoretical distributions was noted.

The double exponential headway distribution model

Schuhl[5] proposed a headway distribution in which vehicles travelling along a highway could be considered to be composed of two types, firstly those who were unable to overtake and were restrained in their driving performance and secondly those drivers who were unrestrained by other vehicles on the highway.

Drivers who are restrained by the action of the driver in front can approach to within a minimum time headway of e and their cumulative headway distribution may be represented by

$$\text{probability (headway} \geq t) = L \exp\left(-(t-e)/(\bar{t}_1 - e)\right) \qquad t \geq e \qquad (12.7)$$

$$= L \qquad\qquad t \leq e$$

where L is the proportion of restrained vehicles in the traffic stream and $\bar{t}_1$ is the mean headway between restrained vehicles.

Similarly drivers who are not restrained by the vehicle in front have no limitation on the minimum headway, as they are able to overtake, and their cumulative headway distribution may be represented by

$$\text{probability (headways} \geq t) = (1 - L) \exp\left(-t/\bar{t}_2\right) \qquad t \geq 0 \qquad (12.8)$$

where $\bar{t}_2$ is the mean headway between unrestrained vehicles.

On the highway both restrained and unrestrained vehicles are present and so the observed headway distribution is represented by the sum of equations 12.7 and 12.8

$$\text{probability (headways} \geq t) = L \exp\left(-(t-e)/(\bar{t}_1 - e)\right)$$

$$+ (1 - L) \exp\left(-t/\bar{t}_2\right) \quad t \geq e \qquad (12.9)$$

Some of the earliest research into the fit of the double exponential distribution to observed headways on two-lane urban streets, was carried out by Kell[7]. The theoretical cumulative headway distribution chosen for fitting to observed values was

$$\text{probability (headway} \geq t) = \exp\left(a - t/K_1\right) + \exp\left(c - t/K_2\right)$$

and the relationships of the parameters to the traffic volume were found to be

$$K_1 = 4827.9/V^{1.024}$$

$$a = -0.046 - 0.000448\,V$$

$$K_2 = 2.659 - 0.0012\,V$$

$$C = \exp(-10.503 + 2.829 \ln V - 0.173\,(\ln V)^2) - 2$$

Similarly Grecco and Sword[7] investigated the headway distribution on a $2\frac{1}{2}$ mile section of US52, a 2 lane 2 way urban bypass around Lafayette, Indiana. They considered that the distribution parameters proposed by Kell were too cumbersome and proposed the following relationships.

Average time headway between restrained vehicles 2.5 s, average time headway between unrestrained vehicles

$$24 - 1.22\ \frac{\text{lane volume}}{100}$$

minimum time headway between restrained vehicles 1.0 s, and proportion of restrained vehicles

$$0.115 \times \frac{\text{lane volume}}{100}$$

The author has carried out observations of headway distributions in the West Riding of Yorkshire and has reported[8] the following relationship between traffic volume and the parameters of the double exponential distribution.

For one-way highways with two traffic lanes, the proportion of restrained vehicles was given by 0.00158 volume − 1.04222 when the volume was between 660 and 1295 vehicles/hour.

For two-way highways with a single traffic lane in each direction and with limited opportunities to overtake by entering the opposing traffic lane, the proportion of restrained vehicles was given by 0.00146 volume − 0.52985 when the volume was between 380 and 790 vehicles/hour.

The fitting of a double exponential distribution to headway data

The use of this double exponential distribution may be demonstrated by an analysis of headways on a highway where traffic is experiencing some degree of congestion. Normally this can be expected to occur on a two-lane two-way highway when the traffic volume in each direction exceeds approximately 600 veh/h or on a two-lane one-way highway when the traffic volume exceeds approximately 1000 veh/h.

The observed headways given in table 12.4 will be used to illustrate the fitting of the double exponential distribution using a graphical technique. Observed headway frequency, cumulative frequency and percentage cumulative frequency distributions are given in table 12.5

The distinctive graphical plot on semi-log paper of the cumulative headway distribution when the traffic flow is partly restrained is illustrated in figure 12.3, which should be compared with figure 12.2 (p. 104) where the cumulative headway distribution for free-flowing traffic is plotted.

TABLE 12.5   Observed headways

Headway class (seconds)	Observed frequency	Number of headways greater than lower class limits	Percentage of headways greater than lower class limit
0–0.9	57	410	100.0
1–1.9	99	353	86.1
2–2.9	69	254	62.0
3–3.9	58	185	45.3
4–4.9	44	127	31.1
5–5.9	25	83	20.4
6–6.9	14	58	14.4
7–7.9	13	44	11.0
8–8.9	11	31	7.8
9–9.9	6	20	5.1
10–10.9	3	14	3.7
11–11.9	3	11	2.9
12–12.9	3	8	2.2
13–13.9	0	5	1.5
14–14.9	1	5	1.5
15–15.9	4	4	1.2

The graphical fitting of the double exponential curve may be described by reference to the form of the theoretical curve shown in figure 12.4. The theoretical cumulative headway distribution for both free flowing and restrained vehicles is shown by a solid curve and, if the observed cumulative headway distribution is plotted, it will approximate to this line if the underlying theoretical concept is correct. The straight line portion of this curve represents the headway distribution of free-flowing vehicles as the effect of restrained vehicles is negligible at the larger values of headway. This portion of the curve may then be represented by equation 12.8. If the straight line portion of the curve is extended to the vertical axis, as shown by a broken line, the percentage of headways greater than t represents the value of $100(1 - L)$ so allowing L to be determined.

If a point is taken on the line and a value of the probability of a headway $\geqslant t$ and the corresponding value of t are substituted in equation 12.8 then the value of $\bar{t}_2$ may be found.

The vertical difference between the straight line and the cumulative curve represents the headway distribution of restrained drivers. Since this difference represents equation 12.6 it is again a straight line. The value of t, when the proportion of restrained vehicles is L, on this straight line gives e. A point may then be selected on the line and a value of the probability and the corresponding value of t substituted in equation 12.7 to give t_1.

Using the parameters $L, e, \bar{t}_1$ and $\bar{t}_2$ the theoretical cumulative headway distribution may be calculated using equation 12.9 and plotted. For instance the theoretical

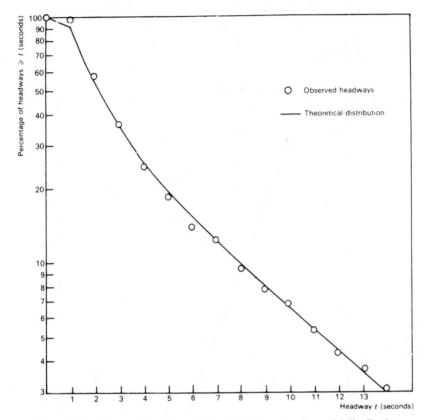

Figure 12.3   Observed headways and the fitted double exponential distribution

values were obtained using the least squares technique and the computer program developed by the author RJSE (10) giving the following values of the parameters

L 0.66

t_1 3.0 seconds

t_2 4.4 seconds

It is interesting to note that at the traffic volume sampled, approximately 1000 veh/h, considerably more than 50 per cent of drivers appear to be restrained by the preceding vehicles. Also at a time headway in the region of 7 seconds drivers appear to cease to be influenced by the vehicle in front.

The observed and the theoretical headway frequencies obtained from the double exponential distribution using the derived parameters are compared in figure 12.5. It can be seen that once again the exponential distribution gives too high a frequency of headways in the class 0-1 second when compared with the observed frequency.

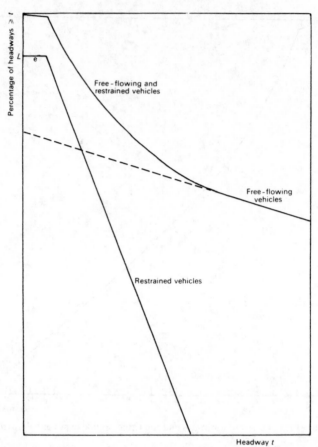

Figure 12.4 The theoretical cumulative headway distribution for free flowing and restrained vehicles in a traffic stream

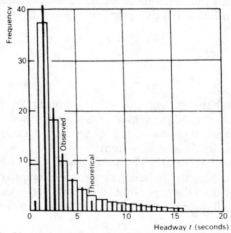

Figure 12.5 Observed and theoretical headway frequencies for congested flow

References

1. F. A. Haight, B. F. Whisler and W. W. Mosher, New statistical method for describing highway distribution of cars, *Proc. Highw. Res. Bd*, **40** (1961) 557-64
2. J.P. Kinzer, Applications of the theory of probability to problems of highway traffic, *Proc. Inst. Traffic Engrs* (1934), 118-23
3. W. F. Adams, Road traffic considered as a random series, *J. Instn Civ. Engrs* (1936), 121-30
4. A. J: Miller, A queueing model of road traffic flow, *J. R. Statist. Soc.*, **23B** (1961), 1
5. A. Schuhl, The probability theory applied to distribution of vehicles on two-lane highways, *Poisson and Traffic*, The Eno Foundation (1955), pp 59-75
6. J. H. Kell, Analysing vehicular delay at intersections through simulation, *Highw. Res. Bd Bull.*, 1 (1962), 28-9
7. W. L. Grecco and E. C. Sword, Predictions of parameters for Schuhl's headway distribution, *Traff. Engng* (Feb. 1968), 36-8
8. R. J. Salter, A simulation of traffic flow at priority intersections, *J. Instn Munic. Engrs*, 98 (Oct. 1971), 10, 239-48

Problems

1. Select the correct completion of the following statements.

(a) The time headway distribution between successive highway vehicles may be obtained
(i)   from an aerial photograph of a highway,
or (ii) by an observer equipped with a stopwatch standing at the side of the highway.

(b) The counting distribution as applied to highway traffic flow indicates
(i)   the distribution of the numbers of vehicles arriving at a point on the highway during fixed periods of time,
or (ii) the distribution of time intervals between the arrival of vehicles,
or (iii) neither

(c) Gap distributions are normally observed by highway traffic researchers because
(i)   the counting distributions require a considerable time to collect adequate data,
(ii) the gap distribution may be easily obtained by the use of aerial photographs.

(d) The time headway distribution is of importance to a study of highway traffic flow because
(i)   the probability of observing a headway increases as the size of the headway decreases,
or (ii) the inverse of the mean time headway is the rate of traffic flow.
(e) The negative exponential distribution represents
(i)   the cumulative headway distribution of freely flowing highway traffic,
or (ii) the cumulative headway distribution of heavily congested highway traffic,
or (iii) the counting distribution of vehicle arrivals on a traffic signal approach.

2. The headway distribution on a two way urban highway is given in table 12.6.
(a) Assume the traffic is free flowing and graphically fit a theoretical distribution to the observed values. Calculate the theoretical frequencies.

TABLE 12.6

Headway class (second)	Observed frequency
0–0.9	19
1–1.9	67
2–2.9	58
3–3.9	29
4–4.9	26
5–5.9	14
6–6.9	17
7–7.9	7
8–8.9	9
9–9.9	6
10–10.9	5
11–11.9	8
12–12.9	4
13–13.9	4
14–14.9	4
15–15.9	3
16–16.9	3
17–17.9	0
18–18.9	2
19–19.9	1

(b) Use a histogram to show the closeness of fit of the theoretical and observed distributions and state what conclusions may be drawn from this comparison.

3. The number of vehicles arriving in two minute intervals at a check point on a highway is given in table 12.7.

TABLE 12.7

No. of vehicles arriving in 2 minute intervals	No. of times this number of vehicles was observed
8	4
7	1
6	4
5	9
4	10
3	11
2	12
1	11
0	1

Show that the traffic flow on this highway is random.

4. The traffic flow on a highway is composed of free flowing and following vehicles. The free flowing vehicles have a mean time headway of 8.0 seconds while the following vehicles do not approach closer to the one in front than a time headway of 0.5 second. When the traffic volume was 1000 vehicles per hour it was noted that there were 25 per cent of free flowing and 75 per cent following vehicles.

Calculate:

(a) the proportion of vehicles with a time headway greater than 6 seconds;
(b) the time headway at which only 5 per cent of vehicles are restrained;
(c) the proportion of following vehicles which have a headway greater than 0.25 second.

Solutions

1. (a) The time headway distribution between successive highway vehicles may be obtained by an observer equipped with a stopwatch standing at the side of the highway.

(b) The counting distribution as applied to highway traffic flow indicates the distribution of the numbers of vehicles arriving at a point on the highway during fixed periods of time.

TABLE 12.8

Class (seconds)	Observed frequency	Observed cumulative frequency	Percentage observed cumulative frequency	Percentage theoretical cumulative frequency	Theoretical cumulative frequency	Theoretical frequency
0–0.9	19	286	100.0	100.0	286	56
1–1.9	67	267	93.4	80.3	230	46
2–2.9	58	200	69.9	64.4	184	36
3–3.9	29	142	49.7	51.7	148	29
4–4.9	26	113	39.5	41.5	119	23
5–5.9	14	87	30.4	36.4	104	19
6–6.9	17	73	25.5	26.7	76	15
7–7.9	7	56	19.6	21.4	61	12
8–8.9	9	49	17.1	17.4	50	10
9–9.9	6	40	14.0	14.0	40	8
10–10.9	5	34	11.9	11.2	32	6
11–11.9	8	29	10.1	9.0	26	5
12–12.9	4	21	7.3	7.2	21	4
13–13 9	4	17	5.9	5.8	17	3
14–14.9	4	13	4.5	4.6	13	3
15–15.9	3	9	3.1	3.7	11	2
16–16.9	3	6	2.9	3.0	9	2
17–17.9	0	3	1.0	2.4	7	1
18–18.9	2	3	1.0	1.9	6	1
19–19.9	1	1	–	1.5	4	0

(c) Gap distributions are normally observed by highway traffic researchers because the counting distributions require a considerable time to collect adequate data.

(d) The time headway distribution is of importance to a study of highway traffic flow because the inverse of the mean time headway is the rate of traffic flow.

(e) The negative exponential distribution represents the cumulative headway distribution of freely flowing highway traffic.

2. To graphically fit a theoretical distribution to the observed headway distribution the percentage cumulative headway distribution will be calculated. If the traffic is free flowing then the arrival of vehicles at a point on the highway will be random and the negative exponential distribution will fit the observed cumulative headway distribution. This can be demonstrated by plotting the observed percentage cumulative distribution on semi-log paper. The relationship will be linear and the best straight line can be fitted to the data to obtain the theoretical cumulative distribution from the straight line. The theoretical cumulative distribution can be converted to a cumulative distribution and the differences between successive values will give the theoretical frequencies.

The observed and theoretical frequencies are compared with the help of a histogram. It can be seen that the first class is a particularly poor fit as can be expected with the exponential distribution in which the probability of a headway increases as the headway size decreases. This does not occur on the highway where there is a minimum headway between vehicles unless these vehicles are overtaking.

3. The calculation can be carried out in tabular form as shown in table 12.9.

TABLE 12.9

(1) No. of vehicles per 2 minute interval	(2) No. of times observed (f_0)	(3) $(1) \times (2)$	(4) Theoretical probability*	(5) $\Sigma (2) \times 4$ Theoretical frequency (ft)	(6) x^2 $\dfrac{(f_0 - f_t)^2}{f_t}$
8	4	32	0.015	0.95 ⎫	
7	1	7	0.036	2.27 ⎬	0.18
6	4	24	0.073	4.60 ⎭	
5	9	45	0.128	8.06	0.11
4	10	40	0.187	11.78	0.27
3	11	33	0.218	13.78	0.54
2	12	24	0.191	12.03	0.00
1	11	11	0.110	6.99 ⎫	
0	1	0	0.032	2.02 ⎭	1.04
	$\Sigma 63$	$\Sigma 216$			$\Sigma 2.14$

*Probability (n vehicles arriving per 2 minute interval)

$$= \frac{\exp(-m)m^n}{n!}$$

where $m = \Sigma((1) \times (2))/\Sigma(2) = 3.43$ vehicles.

The number of degrees of freedom is equal to the number of groups (6) minus the constraints (the mean number of vehicles arriving per 2 minute interval, and the total number of vehicle arrivals).

From tables of chi-squared it can be seen that the value of chi-squared for 4 degrees of freedom at the 5 per cent level of significance is 9.49. There is thus no significant difference at the 5 per cent level of significance between the observed and the theoretical distributions indicating that vehicle arrivals are random.

4. The cumulative headway distribution of the freely flowing vehicles in the stream is given by

$$\text{probability } (h \geqslant t) = (1 - L) \exp(-t/\bar{t}_2)$$

while the cumulative headway distribution of the following vehicles in the stream is given by

$$\text{probability } (h \geqslant t) = L \exp(-(t - e)/(\bar{t}_1 - e))$$

From the information given in the question

$$\text{traffic volume} = \frac{3600}{L\bar{t}_1 + (1 - L)\bar{t}_2}$$

$$\bar{t}_2 = 8.0 \text{ seconds}$$

$$L = 0.75$$

Hence

$$1000 = \frac{3600}{0.75t_1 + 0.25 \times 8.0}$$

that is

$$\bar{t}_1 = 2.1 \text{ seconds}$$

(a) The proportion of vehicles, both free flowing and following, with a time headway greater than 6 seconds is given by

$$\text{probability } (h \geqslant 6) = (1 - 0.75) \exp(-6/8) + 0.75 \exp(-(6 - 0.5)/(2.1 - 0.5))$$
$$= 0.25 \times 0.4724 + 0.75 \times 0.1385$$
$$= 0.1181 + 0.2389$$
$$= 0.3570$$

(b) The probability of a headway relating to a following vehicle may be calculated from

$$\text{probability } (h \geqslant t) = L \exp(-(t - e)/(\bar{t}_1 - e))$$

In this case the equation is solved to obtain t when the probability is 0.05

$$0.05 = 0.75 \exp(-(t - 0.5)/(2.1 - 0.5))$$
$$0.067 = \exp(-(t - 0.5)/1.6)$$

Taking natural logarithms of both sides of the equation

$$-2.7031 = -\frac{t - 0.5}{2.7}$$

$t = 4.8$ seconds.

(c) As all following vehicles have a headway greater than 0.5 second then all following vehicles have a headway greater than 0.25 second. The proportion of following vehicles is 0.75 and so the proportion of following vehicles with a headway greater than 0.25 second is 0.75.

13

The relationship between speed, flow and density of a highway traffic stream

Theoretical relationships between speed, flow and density

When considering the flow of traffic along a highway three descriptors are of considerable significance. They are the speed and the density or concentration, which describe the quality of service experienced by the stream; and the flow or volume, which measures the quantity of the stream and the demand on the highway facility.

The speed is the space mean speed; the density or concentration is the number of vehicles per unit length of highway and the flow is the number of vehicles passing a given point on the highway per unit time.

The relationship between these parameters of the flow may be derived as follows. Consider a short section of highway of length L in which N vehicles pass a point in the section during a time interval T, all the vehicles travelling in the same direction.

The volume flowing $Q = N/T$

$$\text{The density } D = \frac{\text{average no. of vehicles travelling over } L}{L}$$

The average number of vehicles travelling over L is given by

$$\frac{\sum_{i=1}^{N} t_i}{T}$$

where t is the time of travel of the ith vehicle over the length L; then

$$D = \frac{\displaystyle\sum_{i=1}^{N} t_i}{T} \Bigg/ L = \frac{\dfrac{N}{T}}{\dfrac{L}{\dfrac{1}{N}\displaystyle\sum_{i=1}^{N} t_i}}$$

or,

$$\text{density} = \frac{\text{flow}}{\text{space mean speed}} \tag{13.1}$$

Numerous observations have been carried out to determine the relationship between any two of these parameters for, with one relationship established, the relationship between the three parameters is determined. Usually the experimenters have been interested in the relationship between speed and flow because of a desire to estimate the optimum speed for maximum flow.

Greenshields[1] is one of the earliest reported researchers in this field and in a study of rural roads in Ohio he found a linear relationship between speed and density of the form

$$\overline{V}_s = \overline{V}_f - \left(\frac{\overline{V}_f}{D_j}\right) D \tag{13.2}$$

where $\overline{V}_s$ is the space mean speed,
        $\overline{V}_f$ is the space mean speed for free flow conditions,
        D_j is the jam density.

With this relationship determined the flow density relationship can be obtained by substitution of equation (13.1) in equation (13.2) to give

$$Q = \overline{V}_f D - \frac{\overline{V}_f}{D_j} D^2 \tag{13.3}$$

and similarly the relationship between flow and speed may be obtained as

$$Q = D_j \overline{V}_s - \frac{D_j}{\overline{V}_f} \overline{V}_s^2 \tag{13.4}$$

The density and speed at which flow is a maximum can be obtained by differentiating equations 13.3 and 13.4 with respect to density and speed. To obtain the density when flow is a maximum, from equation 13.3

$$\frac{dQ}{dD} = \overline{V}_t - \left(2 \times \frac{\overline{V}_f}{D_j} D\right) = 0 \text{ for a maximum value}$$

$$D = D_{\max} = \frac{D_j}{2} \tag{13.5}$$

To obtain the speed when flow is a maximum, from equation 13.4

$$\frac{dQ}{d\bar{V}_s} = D_j - \left(2 \times \frac{D_j}{\bar{V}_f} \; \bar{V}_s\right)$$

$$\bar{V}_s = \bar{V}_{max} = \frac{\bar{V}_f}{2} \tag{13.6}$$

Substituting these maximum values in equation 13.1 gives $\bar{Q}_{max}$

$$\bar{Q}_{max} = D_{max} \bar{V}_{max} = D_j \bar{V}_f/4 \tag{13.7}$$

Observations obtained by Greenshields gave the following values

$\bar{V}_f = 74$ km/h

$D_j = 121$ vehicles/km

from equation 13.5 $D_{max} = 61$ vehicles/km
from equation 13.6 $\bar{V}_{max} = 37$ km/h
from equation 13.7 $Q_{max} = 2239$ vehicles/h.

Figures 13.1, 13.2 and 13.3 show the relationships between speed and flow, density and speed and flow and concentration respectively.

A considerable number of relationships have been proposed between speed and density and the fit of some of these hypotheses to observed data has been investigated by Drake, Schofer and May[2]. Each hypothesis then affects the relationship between speed and flow and also between density and flow.

Greenberg[3] observed traffic flow in the north tube of the Lincoln Tunnel, New York City. He assumed that high density traffic behaved in a similar manner to a continuous fluid for which the equation of motion is

$$\frac{d\bar{V}_s}{dT} = - \frac{C^2}{D} \times \frac{\partial D}{\partial L}$$

where C is a constant and the remaining symbols are as previously defined. Then

$$\frac{d\bar{V}_s}{dT} = \frac{\partial \bar{V}_s}{\partial L} \times \frac{dL}{dT} + \frac{\partial \bar{V}_s}{\partial T} \times \frac{dT}{dT}$$

$$= \frac{\partial \bar{V}_s}{\partial L} \times \bar{V}_s + \frac{\partial \bar{V}_s}{\partial T}$$

or

$$\frac{\partial \bar{V}_s}{\partial L} \times \bar{V}_s + \frac{\partial \bar{V}_s}{\partial T} + \frac{C^2}{D} \times \frac{\partial D}{\partial L} = 0$$

From the equation of continuity of flow

$$\frac{\partial D}{\partial T} + \frac{\partial Q}{\partial L} = 0$$

and

$$Q = \bar{V}_s D$$

then

$$\bar{V}_s = C \ln (D_j/D)$$

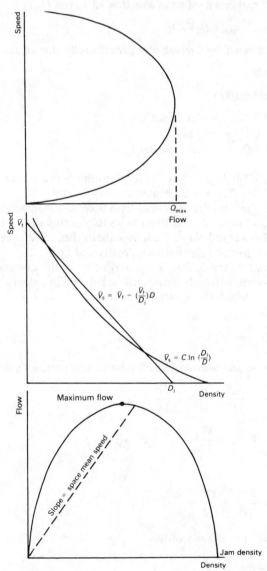

Figure 13.1, 2, 3   The relationship between speed, flow and density for highway traffic flow

Using equation 13.1 gives

$$Q = CD \ln \frac{D_j}{D} \tag{13.8}$$

and

$$D = \frac{CD}{V_s} \ln \frac{D_j}{D} \tag{13.9}$$

It is interesting to note that the Greenberg model does not give $\overline{V}_s = \overline{V}_f$. When $D = 0$ the boundary conditions are:

1. When $Q = 0$, $\underline{D} = 0$ or $D = D_j$.
2. When $D = 0$, $\underline{V}_s = \infty$.
3. When $D = D_j$, $V_s = 0$.

The relationship between flow and density illustrated in figure 13.3 has been referred to as the fundamental diagram of traffic[4]. Its form has received considerable attention from those interested in highway flow and it can be described as follows.

The flow is obviously zero when the density is zero and at the jam density the flow may also be assumed to be zero. Between these limits the flow must rise to at least one maximum, often referred to as maximum capacity, to give a shape of the approximate form shown in figure 13.3. At any point on this curve the slope of the line joining that point to the origin is the space speed. The slope is obviously greatest at the origin and decreases to zero at the jam density.

Lighthill and Whitham[5] using a fluid-flow analogy have shown that the speed of waves causing continuous changes of volume through vehicular flow is given by dQ/dD which is the slope of the fundamental diagram, that is

$$V_w = \frac{dQ}{dD}$$

where V_w is the speed of the wave. From equation 13.1

$$V_w = \frac{d(\overline{V}_s D)}{dD} = \overline{V}_s + D \frac{d\overline{V}_s}{dD} \tag{13.10}$$

The space mean speed decreases with increasing density so that $d\overline{V}_s/dD$ is negative. Hence the wave speed is less than that of the traffic stream.

At low densities the second term of equation (13.10) approaches zero and wave velocity approaches the speed of the stream. At the maximum flow the wave is stationary relative to the road since dQ/dD is zero and at higher densities the waves move backwards relative to the stream.

When movement is viewed relative to the road, the wave moves forward at densities less than that of the maximum flow while the wave moves backwards at densities greater than that of the maximum flow.

Changes in density in a traffic stream will result in the production of differing waves travelling at differing velocities through the stream. An example of this could be a length of highway with a heavily trafficked entrance ramp. The wave in the low density length of highway downstream of the entrance ramp will travel forward relative to the highway at a greater speed than the wave in the higher density length of highway upstream of the entrance ramp. When these waves meet, a shock wave will form which will have a velocity

$$V_{sw} = \frac{Q_2 - Q_1}{D_2 - D_1}$$

where Q_1, Q_2 and D_1, D_2 are the respective points on the fundamental diagram for the low and high densities of flow.

A considerable number of observations have been carried out to determine the relationship between speed, flow and density[6,7,8] and the Highway Capacity Manual[9] gives a comprehensive review of the work carried out in the United States.

Observations of the relationship between speed, flow and density

An indication of the relationship between speed, flow and density for a given point on a highway can be obtained by the following observations.

Select a length of highway where the flow is effectively confined to one lane and a range of flow conditions can be expected. By means of a radar speedmeter connected to a graphical recorder or by using a Rustrak or similar recorder take observations of the speeds and time headways of the stream.

The Rustrak recorder is a battery operated device which allows up to four events to be recorded on a moving chart by the depression of one of four push buttons. At least two observers are required, one to mark the entry of a vehicle into a measured speed trap or baseline and the other to mark the exit of the vehicle from the speed trap. If four observers are available more meaningful results can be obtained by taking into account the proportion of commercial vehicles in the flow. In this case two observers would record passenger cars or similar vehicles and two observers would record medium and heavy goods vehicles. Concentration and accuracy are required in this work if confusion between vehicles is to be avoided and in addition some form of time check on the chart is necessary to obtain the relationship between chart distance and real time.

With the time scale of the chart established it is possible to calculate the speed of the vehicles over the length of the baseline by the difference in time between entry and exit. The time headway can similarly be calculated from the difference in time between successive vehicles entering the baseline.

With data obtained from either source the observations should be divided into 5-minute time intervals and the mean values of speed and headway calculated. Note that it is the space mean-speed which is required and not the time mean-speed. The appropriate formula will be

$$\overline{V}_s = \frac{L}{\dfrac{1}{N} \displaystyle\sum_{i=1}^{N} t_i}$$

where t is the time of travel of the ith vehicle over the measured baseline or speed trap length L when the Rustrak event recorder is used, or

$$\bar{V}_s = \cfrac{1}{\dfrac{1}{N} \displaystyle\sum_{i=1}^{N} \dfrac{1}{V_t}}$$

where V_t is the time mean-speed given by the radar speedmeter.

The mean time headway for each 5 minute interval is obtained by observation and the density is then obtained from

D vehicles/km = 3600/(mean time headway $\bar{V}_s$)

where the mean time headway is measured in seconds and $\bar{V}_s$ is the space mean-speed in km/h.

The flow may then be calculated using the relationship given in equation 13.1.

The observations shown in table 13.1 of the traffic flow in the fast lane of the Lincoln Tunnel, New York City by Olcott[11] have been adapted and are used to illustrate the fundamental relationships between density, flow and speed. Observations were made with an event recorder, the speed of vehicles being estimated by time of travel over a measured baseline. Summarised details of the observations are given in table 13.1.

TABLE 13.1

No. of vehicles observed in a 5 min. period	Space mean speed $\bar{V}_s$ (km/h)	Flow Q (veh/h)	Density D (veh/km)	D^2	$\bar{V}_s D$
97	27.0	1164	43.1	1857.6	1163.7
108	25.4	1296	51.0	2601.0	1295.4
104	30.7	1248	40.7	1656.6	1249.5
100	25.6	1200	46.9	2200.0	1200.6
113	34.8	1356	39.0	1521.0	1357.2
116	41.4	1392	33.6	1129.0	1391.0
116	30.2	1392	46.1	2125.2	1392.2
110	40.4	1320	32.7	1069.3	1321.1
115	39.7	1380	34.8	1211.0	1381.6
91	51.2	1092	21.3	453.7	1090.6
	Σ 346.4		Σ 389.2	Σ 15824.4	Σ 12842.9

The fit of this data to the relationship between speed and density proposed by Greenshields and given in equation 13.2 may be shown by estimating V_f and $(\bar{V}_f/D_j)$ using the method of least squares. In this method

$$\Sigma \ \bar{V}_s = n\bar{V}_f + (\bar{V}_f/D_j) \ \Sigma D$$

$$\Sigma \ \bar{V}_s D = \bar{V}_f \ \Sigma D + (\bar{V}_f/D_j) \ \Sigma D^2$$

Substituting the values obtained by summation in table 13.1 gives

$$\overline{V}_f = 71.4 \text{ km/h}$$

$$\frac{\overline{V}_f}{D_j} = -0.94$$

giving

$$D_j = 76 \text{ vehicles/km}$$

The relationship is

$$\overline{V}_s = 71.4 - 0.94D$$

also

$$Q = 71.4D - 0.94D^2$$

and

$$Q = 1.06 \overline{V}_s (71.4 - \overline{V}_s)$$

These relationships are illustrated graphically in figure 13.4 where observed values and the fitted speed/flow, speed/density and flow/density curves are shown.

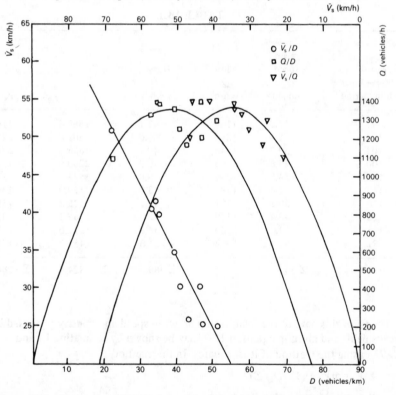

Figure 13.4   Observed and fitted relationships between speed/density, flow/density and speed/flow

From inspection of the graphical plot of flow and space mean-speed, or from differentiation, the maximum flow is found to be 1351 vehicles/h; it occurs when the space mean-speed of the stream is 35.7 km/h.

References

1. B. D. Greenshields, A study of traffic capacity, *Proc. Highw. Res. Bd*, **14** (1934), Pt 1, 448–74
2. J. Drake, J. Schofer and A. May, A statistical analysis of speed–density hypotheses. Vehicular Traffic Science, *Proc. 3rd International Symposium on the Theory of Traffic Flow* (1965), Elsevier, New York (1967)
3. H. Greenberg, An analysis of traffic flow, *Ops. Res.*, 7 (1959), 1, 79–85
4. F. A. Haight, *Mathematical Theories of Traffic Flow*, Academic Press, New York (1963)
5. M. J. Lighthill and G. B. Whitham, A theory of traffic flow on long roads, *Proc. R. Soc.*, **A229** (1959), 317–45
6. Road Research Laboratory, *Research on road traffic*, HMSO, London (1965)
7. J. G. Wardrop, Journey speed and flow in central urban areas, *Traff. Engng Control*, **9** (1968), 11, 528–32
8. A. C. Dick, Speed/flow relationships within an urban area, *Traff. Engng Control*, Oct. (1966), 393–6
9. Highway Capacity Manual, *Highway Research Board Special Report* 87, 1965
10. R. J. Salter, Vehicle speeds and headways at northern end of M1, *Highw. Traff. Engng* (Sept. 1969), 34–6
11. E. A. Olcott, The influence of vehicular speed and spacing on tunnel capacity. Presented at the Informal Seminar in Operations Research, Johns Hopkins University, November 1954

Problems

1. Greenshields proposed a linear relationship between speed and density. Using this relationship it was noted that on a length of highway the free speed V_f was 80 km/h and the jam density D_j was 70 vehicles/km.

(a) What is the maximum flow which could be expected on this highway?
(b) At what speed would it occur?

2. Observations of the speed and flow through a highway tunnel showed that the relationship between speed and density was of the form

$$\bar{V}_s = 35.9 \ln \frac{180}{D}$$

where speeds are measured in km/h and densities in vehicles/km.

(a) What is the jam density on this highway?

3. The diagram below (Figure 13.5) shows the relationship between volume and density for traffic flow on a highway. Four points on the curve are marked A, B, C and D; indicate for the traffic flow situations given below those which could be represented by point(s) A and/or B and/or C and/or D.

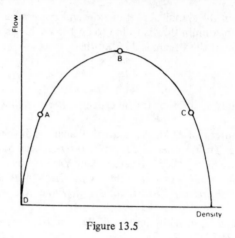

Figure 13.5

(a) Traffic flow conditions where the wave velocity was moving backwards relative to the roadway.

(b) Traffic flow conditions where the wave velocity was less than the space mean-speed of the traffic stream and the wave was moving forward relative to the roadway.

(c) Traffic flow conditions where the wave velocity and the space mean-speed stream velocity were equal.

(d) Congested traffic flow conditions where level of service E prevailed.

(e) Free flow conditions where level of service A prevailed.

Solutions

1. (a) The maximum flow $\bar{Q}_{max}$ is given by

$$D_j \bar{V}_f/4$$

when there is a linear relationship between speed and density, that is

$$\bar{Q}_{max} = \frac{70 \times 80}{4} = 1400 \text{ veh/h}$$

(b) The speed $\bar{V}_{max}$ of which the flow is a maximum is given by

$$\frac{\bar{V}_f}{2} = 40 \text{ km/h}$$

2. The jam density D_j on a highway is given by

$$\bar{V}_s = C \ln \frac{D_j}{D}$$

Hence in the equation given

$$\overline{V}_s = 35.9 \ln \frac{180}{D}$$

the jam density D_j is 180 veh/km.

3. (a) Traffic flow conditions where the wave velocity was moving backwards relative to the roadway (C).

(b) Traffic flow conditions where the wave velocity was less than the space mean-speed of the traffic stream and the wave was moving forward relative to the roadway (A).

(c) Traffic flow conditions where the wave velocity and the space mean-speed stream velocity were equal (D).

(d) Congested traffic flow conditions where the level of service E prevailed (C).

(e) Free flow conditions where level of service A prevailed (D).

14

The distribution of vehicular speeds in a highway traffic stream

Space mean and time mean-speed

One of the fundamental parameters for describing traffic flow is the speed, either of individual vehicles or of the traffic stream. It is of importance in work connected with the theory of traffic flow because of the fundamental connection between speed, flow and concentration when the movement of a traffic stream is being considered.

In traffic management a knowledge of the speed of vehicles is required for the realistic design of traffic signs, the layout of double white lines and the assessment of realistic speed limits. Geometric design of highways also requires that deceleration and acceleration and sight distances, superelevation and curvature must be related to an assumed design speed.

Because speed measurements are used for many purposes in highway traffic engineering and highway design, several differing definitions of speed are commonly used. Where vehicle speed is used to assess journey times, in connection with transportation studies that are concerned with travel over an area, it is the average journey speed that is important. On the other hand if vehicle speed is being used to investigate the accident potential of a section of highway then it is the spot speed which is required. The difference between these two types of speed can be illustrated by a simple example.

Five vehicles travel at 40, 50, 60, 70 and 80 km/h over a distance of 1 km. The mean of the spot speeds is 60 km/h. The mean travel time of the five vehicles is however 0.018 h so that the mean journey speed is 55.6 km/h. When speeds are measured at one point in space over a period it is the time mean-speed that is obtained. These individual speeds are often referred to as spot speeds. It is defined as

$$\bar{V}_t = \frac{\Sigma V_t}{n}$$

where $\overline{V}_t$ is the time mean speed,

V_t are the individual speeds in time,

n is the number of observations.

Where speeds are averaged over space as is the case when the mean journey speed is calculated then it is the space mean-speed that is calculated, that is

$$\overline{V}_s = \frac{n}{\Sigma \, 1/\overline{V}_t}$$

where $\overline{V}_s$ is the space mean-speed,

n is the number of observations.

Wardrop[1] has shown that time and space-mean speeds are connected by the relationship

$$\overline{V}_t = \overline{V}_s + \frac{\sigma_s^2}{\overline{V}_s}$$

where σ_s is the standard deviation of $\overline{V}_s$.

Speed measurements

Most of the speed measurements made by highway traffic engineers produce a time mean-speed because they are obtained by the use of radar speedmeters or timing devices using short baselines. If in the latter case however the mean travel time is used to calculate the mean speed then it is the space mean-speed that is obtained. Speed measurements derived from successive aerial photographs can however be used to derive space mean-speed distributions directly by noting the travel distance of vehicles between successive exposures.

The simplest and cheapest method of obtaining speed data is to measure the time of travel over a measured distance or baseline. The accuracy of the method depends on the method of timing and where a stopwatch is used it is often assumed that a skilled observer can read to 0.2 s but difficulties of a constant reaction time often mean that the level of accuracy is approximately 0.5 s. The Traffic Engineering Handbook[2] suggests baseline lengths of

27 m where stream speed is below 40 km/h
54 m where stream speed is between 40 and 70 km/h
81 m where stream speed is greater than 70 km/h.

Some of the observer errors associated with timing over a measured length may be overcome by the use of the enoscope, which bends the line of sight of the observer by means of a mirror and eliminates parallax error.

Where details of journey times are required then the time of travel over distances of 1 to 2 km may be obtained by the use of observers stationed at each end of this extended baseline and having synchronised stopwatches. The time of arrival of each vehicle into the measured length and the time of its departure are recorded together with some portion of the registration number. Subsequent comparison of times and numbers allows the journey times to be calculated. It is difficult for one observer to read and record more than one registration number every 5 s, even when only a portion of the registration number is recorded. For this reason only registration numbers ending with given digits are recorded by both observers according to the sample size required.

A more realistic form of device for use in present day traffic conditions makes use of two pneumatic tubes attached to the road surface over a short baseline. Most of these instruments are transistorised allowing operation from portable power sources and some allow sampling of vehicle speeds to be carried out, a necessity in heavy traffic flow conditions. As in any observations sampling of the traffic flow should be carried out with care to avoid obtaining a biased sample. Care should be exercised to avoid selecting the first vehicle in a platoon and conversely selection of free flowing single vehicles will tend to bias the resulting speed distribution towards the faster vehicles. Where speed observations are not classified into vehicle type then the combined observations should be in proportion to the traffic composition.

A speed measuring device that does not rely on timing the passage of a vehicle over a base line is the radar speedmeter. These devices operate on the fundamental principle that a radio wave reflected from a moving target has its frequency changed in proportion to the speed of the moving object. Radar meters have been extensively employed both for police and for research purposes. In these meters the radar beam may be transmitted directly along the highway making it necessary to site the meter on a curve or on the central reservation or, as is more usual in the newer models, a beam is transmitted at an angle across the carriageway.

A frequent choice of police forces and traffic engineering departments is the Marconi Portable Electronic Traffic Analyser. This apparatus has an aerial array in which the energy from the transmitter is beamed at an angle of 20° to the main axis of the meter. This means that if the instrument is aimed straight down the road then the radio waves cross the paths of vehicles making it possible to distinguish between individual vehicles.

For police purposes the passage of a vehicle through the beam occurs so quickly that accurate observation of the meter may be difficult. To overcome this difficulty a hold device may be used which causes the meter to remain at the indicated value for a predetermined short time. When used for traffic engineering purposes the speed of successive vehicles may be estimated without the use of this hold device. A graphical recorder may be coupled to the radar meter to obtain a continuous record of vehicle speeds over a considerable period of time. Accuracy of the meter is ±3.2 km/h at speeds up to 130 km/h. Transmitting frequency is in the band 10 675 to 10 699 MHz with a power consumption of 3.5 A at 12 V. A difficulty encountered in the use of this meter with an automatic recording device is the inability to differentiate between vehicles travelling in opposite directions.

Analysis of speed studies

Because in any speed study a considerable number of speeds are observed, statistical techniques are used to analyse the data obtained. Depending upon the accuracy of the data, the use to which the derived results are to be put and the number of observations obtained, a suitable class interval is chosen.

Table 14.1 shows speed observations obtained on a major traffic route. Individual speeds have been grouped into 4 km/h classes given in column 1—an interval which reduces the data into an easily managed number of classes yet does not hide the

TABLE 14.1

1	2	3	4	5	6	7	8
Speed class (km/h)	Frequency	Percentage frequency	Cumulative frequency	Percentage cumulative frequency	Deviation	(2) × (6)	(2) × (6)^2
44–47.9	1	0.286	1	0.286	−9	−9	81
48–51.9	2	0.571	3	0.857	−8	−16	128
52–55.9	2	0.571	5	1.429	−7	−14	98
56–59.9	4	1.143	9	2.571	−6	−24	144
60–63.9	11	3.143	20	5.714	−5	−55	275
64–67.9	24	6.875	44	12.571	−4	−96	384
68–71.9	40	11.429	84	24.000	−3	−120	360
72–75.9	48	13.714	132	37.714	−2	−96	192
76–79.9	63	18.000	195	55.714	−1	−63	63
80–83.9	40	11.429	235	67.143	0	0	0
84–87.9	34	9.714	269	76.857	1	34	34
88–91.9	29	8.286	298	85.143	2	58	116
92–95.9	25	7.143	323	92.286	3	75	225
96–99.9	13	3.714	336	96.000	4	52	208
100–103.9	5	1.429	341	97.429	5	25	125
104–107.9	3	0.857	344	98.286	6	18	108
108–111.9	1	0.286	345	98.571	7	7	49
112–115.9	2	0.571	347	99.143	8	16	128
116–119.9	2	0.571	349	99.714	9	18	162
120–123.9	1	0.286	350	100.000	10	10	100
	Σ 350					Σ −180	Σ 2980

basic form of the speed distribution. In the selection of class intervals thought should be given to the dial readings when visual observation of the speed is made. Most speeds will be recorded to the nearest dial reading and these form convenient mid-class marks.

The number of observations in each class, or the frequency, is given in column 2 and converted into the percentage in each class by dividing the individual values in column 2 by the sum of column 2. This percentage frequency is given in column 3. The cumulative number of observations or cumulative frequency is given in column 4. This column represents the number of vehicles travelling at a speed greater than the lower class limit. In column 5 the percentage cumulative frequency is given. It is obtained by dividing the value in column 4 by the total number of speeds observed.

It is often assumed that speeds are normally distributed. To test this hypothesis it is necessary to estimate the mean speed and the standard deviation of the observed speeds. Both the mean and standard deviation can be calculated by the use of coding to reduce the arithmetic manipulation necessary. A class is selected which is considered likely to contain the mean speed, although it is not essential that the mean does lie within the class. The number of class deviations from this selected class is given in column 6. The sign of the deviation should be particularly noted.

These deviations are multiplied by the corresponding frequency, given in column 2, and the resulting value is entered in column 7.

The mean speed is then given by

$$\text{mid-class mark of selected class} + \frac{\text{class interval } \Sigma \text{ (column 7)}}{\Sigma \text{ (column 2)}}$$

$$82 - \frac{4.180}{350} = 79.9 \text{ km/h}$$

The standard deviation is given by

$$\text{class interval } \sqrt{\left[\frac{\Sigma \text{ (frequency (deviation)}^2)}{\Sigma \text{ (column 2)}} - \left(\frac{\Sigma \text{ (frequency} \times \text{deviation)}}{\Sigma \text{ (column 2)}}\right)^2\right]}$$

The value of Σ (frequency $\times$ deviation) has already been calculated in column 6 and it is now necessary to calculate the frequency (deviation)^2 for each speed class. These values are given in column 8.

$$4 \sqrt{\left[\frac{2980}{350} - \left(\frac{-180}{350}\right)^2\right]} = 11.6 \text{ km/h}$$

The two parameters of the normal distribution have now been determined and it is possible to calculate the theoretical values of speed frequency in each class by either of two methods. These are:

1. by the use of a table of areas under the normal probability curve;
2. by the use of the probit method as developed by Finney[3].

Using tables of the area under a normal probability curve it is possible to calculate the theoretical frequency assuming a normal distribution.

This is performed in a tabular manner in table 14.2. In column 1 the upper class limits used in the original speed observations of table 14.1 are given. The deviation of these class limits from the previously calculated mean are given in column 2 and then converted into standard deviations from the mean by dividing the value in column 2 by the previously calculated standard deviation. These values are given in column 3. From tables of the normal probability curve, the area under a normal curve between these class limits and the mean can be obtained and these values are given in column 4. The difference between successive values in column 4 gives the theoretical area under the normal curve between the class limits. This is the theoretical probability of a speed lying between the class limits and is given in column 5. When this probability is multiplied by the observed total frequency given in table 14.1 the theoretical frequency is obtained. This is given in column 6.

The second method, which uses the probit analysis approach, is based on the percentage cumulative frequency, given in column 5, table 14.1. This approach is described by Finney[2] and uses the relationship that when speeds are distributed normally the cumulative speed distribution may be written.

Percentage of vehicles travelling at a speed equal to, or less than,

$$V = 1/(\sigma \times 2\pi) \int_{-\infty}^{V} \exp\left(-(V - \bar{V})^2/2\sigma^2\right) dv$$

where $\overline{V}$ is the mean speed,

σ is the standard deviation of the speeds.

A demonstration of the fit of the observed cumulative speed distribution to a cumulative normal distribution may be obtained by plotting the probit of the percentage of vehicles travelling at or less than a certain speed, against the speed upper class limit. Values of probits may be obtained from the suggested reading or can be obtained from figure 14.1. The use of this technique converts a cumulative normal curve into a straight line whose equation is

$$\text{Probit of percentage of vehicles travelling at a speed} < V = 5 + \frac{1}{\sigma}(V - \overline{V})$$

Using the derived values of σ and $\overline{V}$ this gives

$$\text{Probit of percentage of vehicles travelling at a speed} < V = 5 + 0.0862(V - 79.9)$$
$$= 0.0862V - 1.6887$$

$$(14.1)$$

TABLE 14.2

1	2	3	4	5	6	7	8
Upper speed class limit (km/h)	Column 1 minus mean speed	Column 2 divided by standard deviation	Normal area	Probability	Theoretical frequency	Observed frequency	$\frac{((6) - (7))^2}{(6)}$
44	−35.9	−3.10	−0.499				
48	−31.9	−2.75	−0.497	0.002	0.7 ⎫	1 ⎫	
52	−27.9	−2.40	−0.492	0.005	1.8 ⎪	2 ⎪	2.27
56	−23.9	−2.06	−0.480	0.012	4.2 ⎬	2 ⎬	
60	−19.9	−1.72	−0.457	0.023	8.1 ⎭	4 ⎭	
64	−15.9	−1.37	−0.415	0.042	14.7	11	0.93
68	−11.9	−1.025	−0.349	0.066	23.1	24	0.04
72	−7.9	−0.680	−0.252	0.097	33.9	40	1.10
76	−3.9	−0.336	−0.132	0.119	41.9	48	0.89
80	+0.1	0.009	0.004	0.137	48.0	63	4.69
84	+4.1	0.354	0.138	0.134	46.9	40	1.02
88	+8.1	0.70	0.258	0.120	42.0	34	1.52
92	+12.1	1.04	0.351	0.093	32.6	29	0.40
96	+16.1	1.39	0.418	0.067	23.4	25	0.11
100	+20.1	1.74	0.459	0.041	14.3	13	0.12
104	+24.1	2.08	0.481	0.022	7.7 ⎫	5 ⎫	
108	+28.1	2.42	0.492	0.011	3.8 ⎪	3 ⎪	
112	+32.1	2.76	0.497	0.005	1.8 ⎪	1 ⎪	0.01
116	+36.1	3.11	0.499	0.002	0.7 ⎬	2 ⎬	
120	+40.1	3.46	0.500	0.001	0.4 ⎪	2 ⎪	
124	+44.1	3.81	0.500	0.000	0 ⎭	1 ⎭	

$$\Sigma\ 13.10$$

This line is plotted in figure 14.1 together with the observed cumulative frequency distribution obtained from column 5 of table 14.1. The close agreement except at high speeds can be seen.

Using the equation above it is possible to calculate the theoretical probit of vehicles travelling at a speed $> V$, and using either the scale given in figure 14.1 or the tables given by Finney it is possible to calculate the theoretical percentage acceptance. This is done in columns 2 and 3 of table 14.3. The difference between successive values in column 3 gives the percentage frequency, which is given in column 4. When this column is multiplied by the total number of speeds observed

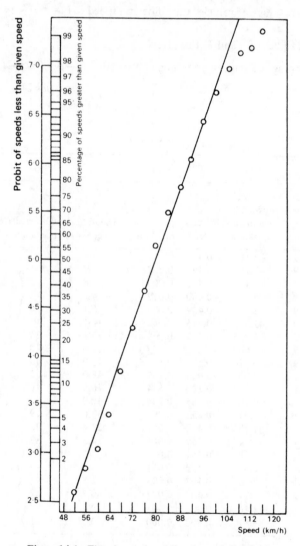

Figure 14.1 The observed and fitted speed distribution

the frequency is obtained and this is given in column 5. The values given in this column differ slightly from those derived in column 6 of table 14.2. This is due to errors involved in the interpolation of values in the tables used in the calculation.

It is often necessary to calculate whether the observed speed frequencies differ significantly from the expected speed frequencies. It is possible to estimate whether the observed and expected results differ significantly by the use of the chi-squared test of significance where chi-squared is given by

$$\chi^2 = \sum \frac{(\text{observed frequency} - \text{expected frequency})^2}{\text{expected frequency}}$$

If χ^2 exceeds the critical value at the 5 per cent level of significance then it is said that there is a significant difference between the observed and the expected values.

The value of chi-squared is calculated in column 8 of table 14.2 and summed at the bottom of the column to give a value of 13.10. It should be noted that where an individual frequency is less than 8 both the observed and the expected frequencies have been combined with other classes.

The critical value of chi-squared which must not be exceeded depends on the number of degrees of freedom. This is given by the number of rows in column 8 (since the greater the number of individual values forming the sum of chi-squared the greater will be its expected value) minus the constraints or the number of para-

TABLE 14.3

1	2	3	4	5
Speed class limit (km/h)	Probit cumulative percentage	Cumulative percentage	Percentage frequency	Frequency
44	2.1031	0.0		
48	2.2502	0.3	0.3	0.1
52	2.5950	0.8	0.5	1.8
56	2.9398	2.0	1.2	4.2
60	3.2746	4.3	2.3	8.1
64	3.6294	8.6	4.3	15.1
68	3.9742	15.3	6.7	23.5
72	4.3190	24.8	9.5	33.3
76	4.6638	36.9	12.1	42.4
80	5.0086	50.3	13.4	46.9
84	5.3534	63.8	13.5	47.3
88	5.6982	75.8	12.0	42.0
92	6.0430	85.2	9.4	32.9
96	6.3878	91.7	6.5	22.8
100	6.7326	95.9	4.2	14.7
104	7.0774	98.1	2.2	7.7
108	7.4222	99.2	1.1	3.9
112	7.7670	99.7	0.5	1.8
116	8.1118	99.9	0.2	0.7
120	8.4566		0.1	0.4
124	8.8014			

meters which were taken from the observed frequency values to calculate the theoretical frequency values. In this case the constraints are the mean speed, the standard deviation of speeds and the total frequency of speeds. This means the number of degrees of freedom of the calculated value of chi-squared is $12 - 3 = 9$.

Tables of chi-squared given in most statistical text books show that the critical value of chi-squared for 9 degrees of freedom at the 5 per cent level of probability is 15.5. The calculated value is 13.10 and so it can be assumed that the observed speed distribution may be represented by a normal distribution.

Having established that speeds are normally distributed it is possible to use the well-known properties of the normal distribution to obtain detailed information on the speed distribution. For example, it is often desirable to calculate the 85 percentile speed and this can be obtained by the use of equation 14.1. The probit of 85 per cent is 6.0364 giving

$$6.0364 = 0.0862V - 1.6887$$

or

$$V(85 \text{ percentile}) = 89.6 \text{ km/h}$$

It is also a property of the normal distribution that 68.27 per cent of all observations will lie within plus or minus one standard deviation of the mean value. This means that approximately two-thirds of all vehicles will be travelling at a speed between 79.9-11.6 km/h and 79.9 + 11.6 km/h, that is between 68.3 and 91.5 km/h. Examination of column 2 of table 14.1 shows that this is correct.

A knowledge of the mathematical form of the speed distribution is also of considerable importance in many theoretical studies of traffic flow and in the simulation of driver behaviour in many highway situations. It has also been suggested[4] that an examination of the speed distribution at a particular location will give an indication of the accident potential of the highway. Where the distribution is normal it is suggested that the accident potential is less than when there is an undue proportion of high or low speeds.

References

1. J. G. Wardrop, Some theoretical aspects of road traffic research, *Proc. Instn Civ. Engrs,* 2 (Pt 1) (1952), 2, 325–62
2. Institute of Traffic Engineers, *Traffic Engineering Handbook* (1965)
3. D. J. Finney, *Probit analysis, a statistical treatment of the sigmoid response curve,* Cambridge University Press (1947)
4. W. C. Taylor, Speed zoning: a theory and its proof, *Traff. Engng,* 35 (1965), 4, pp 17–19, 48, 50–1

Problems

1. Drivers in a vehicle testing programme travel around a race track at a constant speed measured by radar speedmeters and by timing their travel time around the track. The mean speed of all the vehicles was obtained by the two methods and compared.

(a) If all the vehicles travel at the same speed as indicated by their corrected speedometers, will the

(i)   two mean speeds be the same?
(ii) the radar speedmeter mean be higher?
(iii) the radar speedmeter mean be the lower?

(b) If the vehicles travel at differing speeds as indicated by their corrected speedometers, will the

(i)   two mean speeds be the same?
(ii) the radar speedmeter mean be higher?
(iii) the radar speedmeter mean be the lower?

2. The speed distribution on a rural trunk road was noted to be normally distributed with a mean speed of 90 km/h and a standard deviation of 20 km/h. Using the probit technique estimate the 85 percentile speed on the highway.

Solutions

1. When the speeds are measured by a radar speedmeter and averaged it is the time mean-speed V_t that is obtained whereas when speeds are measured from travel time it is the space mean-speed V_s that is calculated. The relationship between the two mean speeds is given by

$$\overline{V}_t = \overline{V}_s + \frac{\sigma_s^2}{\overline{V}_s}$$

(a) In this case all the vehicles travel at the same speed and the standard deviation of the space mean-speeds σ_s is zero so that the two mean speeds are the same.

(b) In this case the vehicles travel at differing speeds and since the standard deviation of speeds is positive the mean speed obtained from the radar speedmeter will be higher than the mean speed obtained from journey times.

2. When speeds can be represented by the cumulative normal distribution then the following relationship may be used.

Probit of percentage of vehicles travelling at a speed $> V = 5 + \frac{1}{\sigma} (V - \overline{V})$

Using the values given

Probit of percentage of vehicles travelling at a speed $> V = 5 + \frac{1}{20} (V - 90)$

The probit of 85 per cent is 6.03, that is

$$6.03 = 5 + \frac{1}{20} (V_{85} - 90)$$

or

$$V_{85} = 110.6 \text{ km/h}$$

15

The macroscopic determination of speed and flow of a highway traffic stream

The relationships considered in chapter 13 between speed, flow and density of a traffic stream considered traffic flow in a microscopic sense in that the speeds of, and headways between, individual vehicles have been used to obtain relationships for the whole traffic stream. An alternative approach to the determination of speed/flow relationships is the use of macroscopic relationships obtained by the observation of stream behaviour.

Speed and flow of a moving stream of vehicles may be obtained by what has become known as the moving car observer method. Wardrop and Charlesworth[1] have described how stream speed and flow may be estimated by travelling in a vehicle against and with the flow. The journey time of the moving observation vehicle is noted, as is the flow of the stream relative to the moving observer. This means that when travelling against the flow the relative flow is given by the number of vehicles met and while travelling with the flow the relative flow is given by the number of vehicles that overtakes the observer minus the number that the observer overtakes. Then

$$Q = \frac{(x + y)}{(t_a + t_w)}$$

and

$$\overline{t} = t_w - y/Q$$

where Q is the flow,
$\overline{t}$ is the mean stream-journey time,
x is the number of vehicles met by the observer, when travelling against the stream,
y is the number of vehicles that overtakes the observer minus number overtaken,

t_a is the journey time against the flow,

t_w is the journey time with the flow.

These relationships are determined as follows. Consider a stream of vehicles moving along a section of road, of length l, so that the average number Q passing through the section per unit time is constant. This stream may be regarded as consisting of flows

Q_1 moving with speed V_1

Q_2 moving with speed V_2

etc.   etc.

Suppose an observer travels with the stream at speed V_w and against the stream at speed V_a, then the flows relative to him, of vehicles with flow Q_1 and speed V_1 are

$$Q_1(V_1 - V_w)/V_1$$

and

$$Q_1(V_1 + V_a)/V_1$$

respectively.

If t_w is the journey time $1/V_w$, t_a is the journey time $1/V_a$ and t_1 is the journey time $1/V_1$, etc., then

$$Q_1(t_a + t_1) = x_1 \quad \text{and} \quad Q_2(t_a + t_2) = x_2, \text{etc.}$$

$$Q_1(t_w - t_1) = y_1 \quad \text{and} \quad Q_2(t_w - t_2) = y_2, \text{etc.}$$

where x_1, x_2, etc., are the number of vehicles travelling at speeds V_1, V_2, etc. met by the observer when travelling against the stream and y_1, y_2, etc., are the number of vehicles overtaken by the observer minus the number overtaking the observer.

Summing over x and y

$$x = x_1 + x_2 + x_3 + \ldots$$

$$y = y_1 + y_2 + y_3 + \ldots$$

$$x = Q_1(t_a + t_1) + Q_2(t_a + t_2) + \ldots$$

$$y = Q_1(t_w - t_1) + Q_2(t_w - t_2) + \ldots$$

$$x = Qt_a + \Sigma(Q_1 t_1 + Q_2 t_2 + \ldots)$$

$$y = Qt_w - \Sigma(Q_1 t_1 + Q_2 t_2 + \ldots)$$

and

$$\bar{t} = \frac{(Q_1 t_1 + Q_2 t_2 + Q_3 t_3 + \ldots)}{Q}$$

so that

$$x = Qt_a + Q\bar{t}$$

$$y = Qt_w - Q\bar{t}$$

or

$$Q = (x + y)/(t_a + t_w)$$

and

$$\bar{t} = t_w - y/Q$$

so allowing the stream speed to be calculated.

The application of this method of measuring speed and flow is illustrated by the following field data, which was obtained to estimate the two-way flow on a highway. To reduce the number of runs required by the observation vehicles, details of the relative flow of both streams are obtained for each run of the observer's vehicle. For this reason, details are given of two relative flows in the data obtained when the observer is travelling eastwards and when he is travelling westwards. These observations are given in table 15.1.

Observer travelling to east　　　　　**TABLE 15.1**

Line	Time of commencement of journey	Journey time	No. of vehicles met	No. of vehicles overtaking observer	No. of vehicles overtaken by observer
1	09.20	2.51	42	1	0
2	09.30	2.58	45	2	0
3	09.40	2.36	47	2	1
4	09.50	3.00	51	2	1
5	10.00	2.42	53	0	0
6	10.10	2.50	53	0	1

Observer travelling to west

Line	Time of commencement of journey	Journey time	No. of vehicles met	No. of vehicles overtaking observer	No. of vehicles overtaken by observer
7	09.25	2.49	34	2	0
8	09.35	2.36	38	2	1
9	09.45	2.73	41	0	0
10	09.55	2.41	31	1	0
11	10.06	2.80	35	0	1
12	10.15	2.48	38	0	1

Length of highway test section 1.6 km.

The information relating to the *highway flow to the east* is abstracted and tabulated in table 15.2.

TABLE 15.2

| Time of commencement of journey | Relative flow rate | | t_a | t_w | Q (veh/min) | $\bar{t}$ (min) | V (km/h) |
	with observer	against observer					
09.20	1			2.51			
09.25		34	2.49		7.0	2.37	41
09.30	2			2.58			
09.35		38	2.36		8.1	2.33	42
09.40	1			2.36			
09.45		41	2.73		8.3	2.24	43
09.50	1		3.00				
09.55		31		2.41	6.3	2.84	34
10.00	0			2.42			
10.05		35	2.89		6.6	2.42	40
10.10	−1			2.50			
10.15		38	2.48		7.4	2.36	41

Similarly, the information relating to the highway flow to the west is abstracted and tabulated in table 15.3.

TABLE 15.3

| Time of commencement of journey | Relative flow rate | | t_a (min) | t_w (min) | Q (veh/min) | $\bar{t}$ (min) | V (km/h) |
	with observer	against observer					
09.20		42	2.51				
09.25	2			2.49	8.8	2.38	41
09.30		45	2.58				
09.35	1			2.36	9.3	2.25	43
09.40		47	2.36				
09.45	0			2.73	9.3	2.62	37
09.50		51	3.00				
09.55	1			2.41	9.6	2.31	42
10.00		53	2.42				
10.05	−1			2.89	9.8	2.79	35
10.10		53	2.50				
10.15	−1			2.48	10.4	2.39	40

If the mean stream speed and flow is required then these individual values may be averaged, to give

stream velocity = 40 km/h

stream flow   = 9.5 veh/min

Reference

1. J. G. Wardrop and G. Charlesworth, A method of estimating speed and flow of traffic from a moving vehicle, *J. Instn Civ. Engrs*, 3 (Pt 2) (1954), 158–71

Problem

In a stream of vehicles 30 per cent of the vehicles travel at a constant speed of 60 km/h, 30 per cent at a constant speed of 80 km/h and the remaining vehicles travel at a constant speed of 100 km/h. An observer travelling at a constant speed of 70 km/h with the stream over a length of 5 km is passed by 17 vehicles more than he passes. When the observer travels against the stream at the same speed and over the same length of highway the number of vehicles met is 303.

(a) What is the mean speed and flow of the traffic stream?
(b) Is the time mean or the space mean-speed obtained by this technique?
(c) How many vehicles travelling at 100 km/h pass the observer, while he travels with the stream?

Solutions

1. (a) The flow Q of a traffic stream may be obtained from

$$\frac{x + y}{t_a + t_w}$$

The time of travel t_a and t_w is 5 km at 70 km/h. Then

$$Q = \frac{303 + 17}{5/70 + 5/70}$$

$$= 2240 \text{ vehicles/h}$$

and

$$\bar{t} = \frac{5}{70} - \frac{17}{2240}$$

$$= 0.0714 - 0.0076$$

$$= 0.0638$$

The mean speed of the stream = 5/0.0638 km/h

$$= 78.3 \text{ km/h}$$

(b) The speed obtained by the moving observer technique is calculated from a journey time and hence it is the space mean-speed that is obtained.

(c) It has been shown that

$$Q_1(t_w - t_1) = y_1$$

where Q_1 is the flow of vehicles with a speed of 100 km/h. In this case it is
0.4 × 2240 or 896 veh/h,

t_w is the travel time of the observer, in this case, 0.071 h,

t_1 is the travel time of the vehicles travelling at 100 km/h, that is 0.05 h,

y_1 is the number of vehicles travelling at 100 km/h that overtake the
observer.

Then

$$896(0.071 - 0.05) = y_1$$

$$= 18 \text{ or } 19 \text{ vehicles}$$

16

Intersections with priority control

Intersections are of the greatest importance in highway design because of their effect on the movement and safety of vehicular traffic flow. The actual place of intersection is determined by siting and design and the act of intersection by regulation and control of the traffic movement. At an intersection a vehicle transfers from the route on which it is travelling to another route, crossing any other traffic streams which flow between it and its destination. To perform this manoeuvre a vehicle may diverge from, merge with, or cross the paths of other vehicles.

Priority control of traffic at junctions is one of the most widely used ways of resolving the conflict between merging and crossing vehicles. The universal adoption of the 'Give Way to traffic on the right' rule at roundabouts together with the use of 'Give Way' and 'Stop' control at junctions has considerably increased the number of occasions at which a driver has to merge or cross a major road traffic stream making use of gaps or lags in one or more conflicting streams.

Both urban and rural motorways of the future will make use of priority control at grade separated junctions where merging vehicles have to enter a major traffic stream and whilst some authorities plan to use traffic signal control for the junctions on their proposed inner ring roads, the relatively low volumes of flow during non-peak hours and the difficulty of dealing with a large proportion of right turning vehicles at traffic signals leads to the roundabout being preferred in many cases.

Types of priority intersections

There are three basic types of major–minor priority junctions. These are: simple junctions, ghost island junctions and junctions with single lane dualling which may be used for crossroads, T-junctions, right–left stagger and left–right stagger junctions. In various forms these junctions may be used on single carriageway roads.

Simple major–minor junctions are used for T-junctions or staggered junctions without any ghost islands in the major road and without any channellising islands in the minor road approach. For junctions in the United Kingdom the Department

146

of Transport[1] state that this type is appropriate for most accesses and minor junctions on single carriageway roads. For new construction they are only recommended when the design flow on the minor road is not expected to have a two-way annual average daily traffic flow exceeding approximately 300 vehicles. For existing junctions the improvement of a simple junction to a ghost island junction should be considered when the minor road two-way annual average daily traffic flow exceeds 500 vehicles or when a right-turning vehicle accident problem exists.

Ghost island junctions have a painted hatched island in the middle of single carriageway roads which provide a diverging lane and a waiting space for right-turning vehicles from the major road into the minor road. Vehicles which turn right from the minor road into the major road are also assisted because in heavy major road flow conditions the minor road vehicle can enter the major road in two stages. The additional construction cost compared with a simple junction is small and is amply justified by the increased safety of the junction.

Single lane dualling at major–minor junctions is achieved by the use of a physical island in the middle of a major single carriageway road. This provides an offside diverging lane for right-turn major road vehicles and a safe waiting area for right-turn vehicles from both the major and minor roads. An important safety feature is the provision of only one through lane in each direction so preventing overtaking and reducing speed in the vicinity of the junction.

On continuous dual carriageway roads major–minor junctions are provided by increasing the width of the central reservation. This allows a diverging lane and a waiting space to be constructed and assists right-turn drivers from the minor road by allowing them to wait within the central reservation before completing this turning movement.

Where major–minor roads intercept to form crossroads then priority control is only appropriate when the minor road flows are low. Crossroads should not be constructed in rural areas or used with dual carriageways or single lane dualling. The staggered junction is preferable with vehicles leaving the minor road turning right and after travelling along the major road turning left into the minor road. This form of staggered junction is recommended rather than the left right staggered junction.

From the point of view of road safety it is recommended[1] that the elimination of lightly trafficked junctions and the collection of these traffic movements at a single junction be carried out wherever possible.

Accidents at priority intersections

Where these merging, diverging and crossing manoeuvres take place a potential, if not an actual, collision may take place between vehicles. The point of a potential or actual collision and the zone of influence around it has been defined as a conflict area.

The number and type of conflict areas may be taken as a measure of the accident potential of a junction, as shown in figures 16.1 and 16.2. It is shown in figure 16.1 that for a four-way priority type junction there are 16 potential crossing conflicts, 8 diverging conflicts and 8 merging conflicts. When this same intersection is converted to a right–left stagger as shown in figure 16.2, then there are only 6 crossing conflicts, 6 diverging conflicts and 6 merging conflicts. While there are objections to such a simplified approach to intersection safety, this method does give an indication of the increase in safety achieved by a staggered crossing.

Intersections may be divided into the following types: (a) priority intersections; (b) signalised intersections; (c) gyratory or rotary intersections; (d) grade-separated intersections.

In the first three types the conflicts between vehicles are resolved by separation in time while in the fourth type separation is used to resolve the main vehicle con-

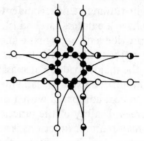

Figure 16.1   Conflict areas at a crossroads

16 crossing conflicts
8 diverging conflicts
8 merging conflicts

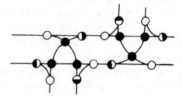

Figure 16.2   Conflict areas at a staggered crossroads

6 crossing conflicts
6 diverging conflicts
6 merging conflicts

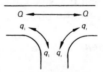

Figure 16.3   Vehicle streams at a three-way intersection

flicts. In the selection of intersection type the following factors should be considered: traffic volumes and delay design speeds; pedestrian movements; cost and availability of land; the accident record of the existing intersection.

An early investigation into accidents at rural three-way junctions was carried out by J. C. Tanner of the Transport and Road Research Laboratory[2] Traffic flows entering or leaving the minor road were divided into flows around the left and right-hand shoulders of the junction respectively as shown in figure 16.3 making the assumption that flows in opposite directions in the same paths were equal.

By means of regression analysis of accident data obtained from 232 rural junctions it was found that an approximate relationship between the annual accident rate around the left and right-hand shoulders of the junction A_i and A_r respectively could be given by the equations

$$A_i = R_1 \sqrt{q_i Q}$$

$$A_r = R_f \sqrt{q_r Q}$$

where q_i, q_r, Q are the August 16-hour flows around the left and right-hand shoulders and on the major road in each direction respectively. For the data studied R_r and R_1 were found to be 4.5×10^{-4} and 7.5×10^{-4}, respectively.

A more recent investigation into junction accident rates has been used to provide data for the Department of Transport COBA program used for the assessment of highway schemes[3].

The approach used in forecasting future accident numbers at new or existing junctions is that the annual number of accidents occurring within 20 m of each junction is estimated from formulae which are of two types, both have the form

Annual number of accidents $= x(f)^y$

where f is a function of traffic flow, and x and y vary according to the type of junction.

The two models are referred to as the cross-product model and the inflow model. In the cross-product model the value of the function f is obtained by multiplying the combined inflow from the two major opposing links by the sum of the inflows on the other one or two minor links. These flows are expressed as thousands of vehicles per annual average day. In the inflow model the value of the function f is the total inflow from all links expressed once again as thousands of vehicles per annual average day.

Junction classification for these models is very coarse and for priority major–minor junctions the designation of junction type depends upon the number of arms of the junction and the nature of the links forming the junction. This classification is shown in table 16.1. For each type of major–minor junction the coefficients x and y vary; these coefficients are given in table 16.2.

TABLE 16.1 Major–minor junction classification for accident estimation

Junction no.	No. of arms	Highest link standard (single or dual)	Formula type
1	3	single	cross-product
2	3	dual	cross-product
3	4	single	inflow
4	4	dual	cross-product

TABLE 16.2   Coefficients for use in accident estimation models

Junction no.	x	y
1	0.195	0.460
2	0.195	0.460
3	0.361	0.440
4	0.240	0.710

A more detailed study of accidents at rural T-junctions has been carried out by Pickering et al.[4]. When considering total junction accidents within 20 m of a junction they found the major–minor flow product model supported previous accident research. The model was of the form,

$$\text{Annual number of accidents} = 0.195 \, (f)^{0.49}$$

where f is the average annual daily traffic expressed as thousands of vehicles

This relationship which was obtained from a study of 299 T-junctions with considerable variation in junction type indicated that the relationships presented in COBA under-predicted accidents by 20 per cent. It was considered that the difference could be due to the COBA relationship having been derived from studies of new or recently modified junctions.

Considerably more detailed results are also presented by Pickering et al. which classified accidents by the movements of the vehicle involved in the accidents and related them to geometric features of these junctions for accidents at distances of between 0 and 20 m and between 20 and 100 m from the junction.

References

1. Department of Transport, Junctions and Accesses: the layout of major/minor junctions, *Departmental Advice Note* TA 20/84, London (1984)
2. M. G. Colgate and J. C. Tanner, Accidents at rural three way junctions, *Road Research Laboratory Report* LR 87, Transport and Road Research Laboratory. Crowthorne (1967)
3. Department of Transport, The COBA program, London (1981)
4. D. Pickering, R. D. Hall and M. Grimmer, Accidents at rural T-junctions, *Research Report* 65, Transport and Road Research Laboratory, Crowthorne (1986)

Problems

1. At a T-junction the estimated annual average daily traffic is 3500 and 4000 vehicles on the major road and 1500 vehicles on the minor road. Calculate the expected annual number of accidents using the COBA model.

2. A road layout has the AADT flows shown below. Show that the combination of these two junctions into a single junction will reduce the expected number of annual accidents.

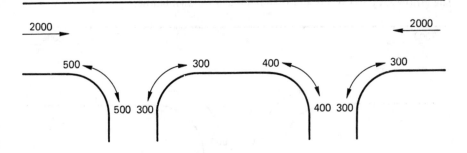

Solutions

1. The COBA model is the cross-product model and the coefficients x and y are 0.195 and 0.460 respectively. The function f is given by $(3.5 + 4)\,1.5$.

Annual number of accidents $= 0.195\,(11.25)^{0.46}$

$$= 1.4$$

2. The inflows at junction A are shown below.

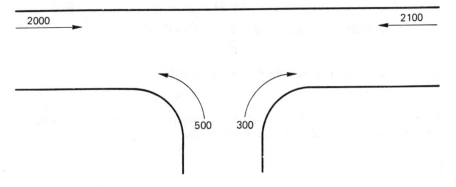

The inflows at junction B are shown below.

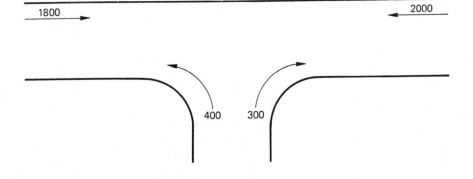

The inflows at junctions A and B combined are shown below.

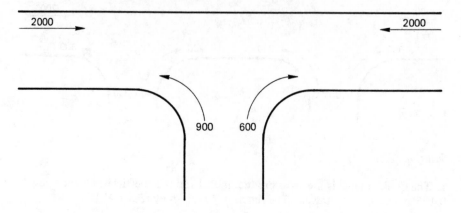

Annual accidents $= x(f)^y$; from table 16.2 x is 0.195 and y is 0.46 and f is the cross-product of the inflows expressed as $AADT/1000$.

$$\text{Annual accidents at junction A} = 0.195(4.1 \times 0.8)^{0.46}$$

$$= 0.34$$

$$\text{Annual accidents at junction B} = 0.195(3.8 \times 0.7)^{0.46}$$

$$= 0.31$$

$$\text{Annual accidents at combined A and B} = 0.195(4 \times 1.5)^{0.46}$$

$$= 0.44$$

It can be seen that the combination of the two junctions reduces the estimated annual accidents from 0.34 + 0.31 or 0.65 to 0.44.

17

Driver reactions at priority intersections

Interaction between traffic streams is an important aspect of highway traffic flow. It occurs when a driver changes traffic lane, merging with or crossing a traffic stream. Probably it takes place most frequently when priority control is used to resolve vehicular conflicts at highway intersections.

When a minor road driver arrives at an intersection he may either enter the major road by driving into a gap in the major road traffic stream or he may reject it as being too small and wait for a subsequent gap. It is important to differentiate when taking observations between lags and gaps. A lag is an unexpired portion of a gap which remains when a minor road driver arrives at the junction of the major and minor roads.

A minor road driver may make only one acceptance decision but he may make many rejections. If every decision of each driver is included in the observations then the resulting gap or lag acceptance distribution will be biased towards the slower driver.

To avoid bias in the observations it is possible to observe only first driver decisions. In this way most of the observed values will be lags and acceptances will be made by drivers from a rolling start. Second, third and subsequent driver decisions will be gap acceptances and will normally be higher in value because of the stationary start.

The usual hypothesis of minor road driver behaviour at priority intersections is the time hypothesis. It is assumed that a gap and lag acceptance judgment is formed as if a minor road driver made a precise estimate of the arrival time of the approaching major road vehicle. The minor road driver is thus able to estimate the time remaining before the approaching vehicle reaches the intersection and merges or crosses on the basis of the time remaining.

Observation of gap and lag acceptance

Lag acceptance behaviour can be studied by observing driver reactions at a priority intersection where the flow on the minor road exceeds 100 veh/h and where the major road traffic stream or streams exceed 400 veh/h.

Observations should be taken of one class of minor road vehicle making one type of turning movement, preferably observing left-turning cars to reduce the time required to complete the observations.

Using a stopwatch the time is noted at which a minor road vehicle arrives at the intersection and also the time at which the next conflicting major road vehicle arrives at the intersection. It is noted also whether the minor road driver accepts the lag and drives out into the major road or rejects the lag and remains in the minor road. After the first driver decision all further actions of the driver are ignored to reduce bias in the observations.

The difference between the two recorded times is the accepted or rejected lag. When observations are taken with a stopwatch it is normally only possible to record lag acceptance to the nearest 0.5 second and so it is convenient to group the observed lag acceptances and rejections into one-second classes.

A set of observations for left-turning passenger cars is shown in table 17.1 Notice that observation was continued until at least 25 observations were obtained in each class.

Column 1 contains the classes that run between zero and 100 per cent acceptance. Columns 2 and 3 give the numbers of observed lags that were rejected and accepted in each class. Column 4 is obtained by dividing column 3 by the sum of columns 2 and 3, the resulting value being expressed as a percentage.

TABLE 17.1   Observed rejected and accepted lags for left-turning vehicles

1	2	3	4
Lag class (s)	Number of observed rejections	Number of observed acceptances	Percentage observed acceptance
0.5–1.4	25	0	0
1.5–2.4	82	2	2
2.5–3.4	56	23	29
3.5–4.4	47	23	33
4.5–5.4	28	41	59
5.5–6.4	17	41	71
6.5–7.4	5	30	86
7.5–8.4	2	48	96
8.5–9.4	0	30	100

If the percentage acceptance is plotted against lag class mark as in figure 17.1 then it can be seen that the curve is approximately of cumulative normal form. This is to be expected in any action that is dependent on human reaction.

It is often necessary in theoretical traffic flow studies and in simulation work to find a mathematical distribution that approximates to an observed phenomenon.

When lag acceptance is of a cumulative normal form then the distribution may be written

Proportion of drivers accepting a lag of t second $= 100/(\sigma\sqrt{2\pi}) \int_{-\infty}^{t} \exp\left(-(t - \bar{t})^2/2\sigma^2\right)$

where $\bar{t}$ and σ are the mean and standard deviation.

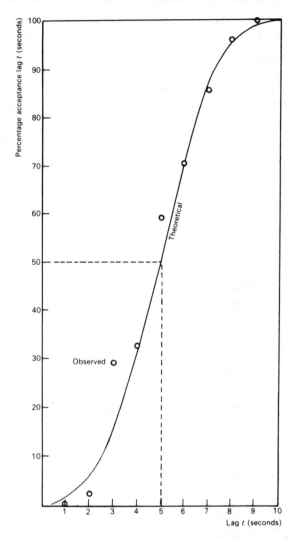

Figure 17.1   Lag acceptance distribution at a priority intersection

A demonstration of the fit of the observed lag acceptance to a normal distribution may be obtained by plotting the probit of the acceptance against the lag class
mark. Values of probits may be obtained from reference 3 (see page 143) or can be
read from figure 17.2. The use of this technique converts a cumulative normal curve
into a straight line and allows the mean and the standard deviation and also the
theoretical values of acceptance to be estimated.

In figure 17.2 the best straight line is drawn through the observed point values.
The mean lag acceptance t is the value when the probit is 5 and the standard devia-

tion of lag acceptance is equal to the reciprocal of the slope of the line. From figure 17.2 these values are

$\bar{t}$ = 5.0 seconds

σ = 1.8 seconds

The theoretical percentage acceptances can also be read from the graph and are given in table 17.2. The theoretical lag acceptance curve and the observed values are shown in figure 17.1 but a statistical test of the closeness of fit can be made by the chi-squared test. Column 8 is calculated from columns 3 and 7. The first three rows have been summed to give an adequate number of acceptances in each group.

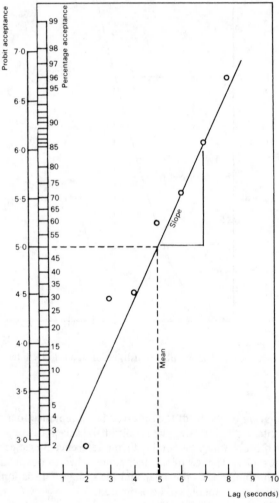

Figure 17.2   Transformation of cumulative normal distribution to a linear relationship using probit acceptance

TABLE 17.2   Theoretical rejected and accepted lags for left-turning vehicles

5	6	7	8
Lag class (seconds)	Percentage theoretical acceptance	Number of theoretical acceptances	Chi-squared
0.5–1.4	2	1	
1.5–2.4	6	5	
2.5–3.4	16	13	1.90
3.5–4.4	32	22	0.05
4.5–5.4	54	37	0.43
5.5–6.4	74	43	0.09
6.5–7.4	88	31	0.03
7.5–8.4	96	48	0
8.5–9.4	99	30	0

Σ 2.50

The summed value of chi-squared has four degrees of freedom: the seven groups less the three constraints; the mean and standard deviation of the lag acceptances; and the total number of decisions.

From statistical tables it can be seen there is not a significant difference at the 5 per cent level between the observed and theoretical distribution.

Many observations of merging performance of vehicles have indicated that a cumulative log-normal distribution of lag acceptance is a better fit than the cumulative normal distribution. This can be tested using the probit transformation by the graphical fitting of a straight line to the observed lag acceptance with the log lag class-mark replacing the lag class-mark. A chi-squared test can then be used to compare the closeness of fit with the cumulative normal distribution.

A different value of lag was used by Raff and Hart[1] in investigations into driver behaviour in New Haven, Connecticut. They defined a critical lag as that lag of which the number of accepted lags shorter than it is equal to the number of rejected lags longer than it.

This critical lag is obtained graphically in figure 17.3 where the critical lag is determined graphically to be 4.6 seconds. This approach is only useful in giving a value of lag acceptance that may be used in traffic studies that use average values of driver performance. It was used by Tsongos and Weiner[2] to compare the effect of darkness on gap acceptance, a subject of considerable interest because of the proportion of peak hour flows that occurs during the hours of darkness.

It has been shown that the critical lag and the mean of the lag acceptance distribution are related in the case where the lag acceptance distribution is normally distributed. This relationship is

$$\text{critical lag} = \bar{t} - \sigma^2 q/2$$

where $\bar{t}$ and σ are the mean and standard deviation of the lag acceptance distribution and q is the flow of vehicles on the major road.

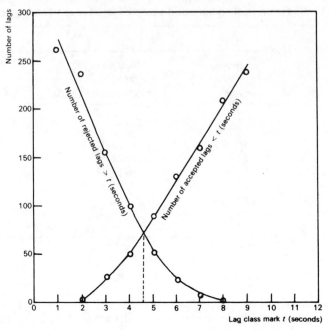

Figure 17.3   Graphical determination of the critical lag

References

1. M. S. Raff and J. W. Hart, *A volume warrant for urban stop signs*, The Eno Foundation for Highway Traffic Control, Saugatuck, Conn., U.S.A. (1950)
2. N. G. Tsongos and S. Weiner, Comparison of day and night gap acceptance probabilities, *Publ. Rds, Wash.*, **35** (1969), 7, 157-65

Problems

1. Observations of driver lag acceptance at an intersection showed that it was normally distributed with a mean of 5.0 s and a standard deviation of 1.5 s.

(a) Is the percentage of drivers requiring a lag greater than 7.0 s, 9 per cent, 15 per cent or 31 per cent?

(b) If it is noted that the mean of the combined lag and gap acceptance distribution is 6.0 s would this indicate

(i)   that drivers entering the major road without stopping are likely to enter the major road more readily than those drivers who have come to a halt before entering?
(ii)   that the second decision gap acceptance of drivers has a lower mean value than first decisions?
(iii)   neither of these?

(c) What is the value of the critical lag at this intersection when the major road flow is 600 veh/h?

Solutions

1. (a) The percentage of drivers accepting a lag greater than t seconds may be obtained from the straight-line relationship between the probit of the percentage of lag acceptance and the time lag, which may be stated as

probit percentage acceptance = $a + b \times$ lag t s

where b = the reciprocal of the standard deviation of lag acceptance. From the value of the mean and standard deviations given and probit tables

$$5 = a + \frac{5}{1.5}$$

and

$$a = 1.67$$

so that

$$\text{probit percentage acceptance} = 1.67 + \frac{\text{lag}}{1.5}$$

When the lag is 7.0 s

$$\text{probit percentage acceptance} = 1.67 + \frac{7.0}{1.5}$$

$$= 6.34$$

From probit tables (or from figure 17.2) percentage acceptance = 91 per cent. The percentage of drivers who require a lag greater than 7.0 s is therefore

$$100 - 91 \text{ per cent} = 9 \text{ per cent}$$

(b) As the combined lag and gap acceptance distribution has a mean of 6.0 s and the lag acceptance distribution has a mean of 5.0 s this indicates that the gap acceptance distribution has a larger value of mean acceptance than the lag acceptance. Hence (1) is correct since drivers who enter the major road without stopping and hence accept lags do so more readily than those who come to a halt.

(c) From the relationship

$$\text{critical lag} = \bar{t} - \sigma^2 q/2$$

$$\bar{t} = 5.0 \text{ s}$$

$$\sigma = 1.5 \text{ s}$$

$$q = 600/3600 \text{ veh/s}$$

$$\text{critical lag} = 5.0 - 1.5^2/6.2$$

$$= 4.8 \text{ s}$$

18

Delays at priority intersections

In contrast to traffic signal-controlled junctions, where saturation flows and resulting junction capacities can be easily calculated, the estimation of the practical capacity of priority type junctions presents considerable difficulties.

At the present time the growth of traffic on a largely unimproved urban highway system has resulted, even in the smaller urban areas, in considerable congestion and delay at intersections during peak hours. These symptoms of traffic growth were reached many years ago in the United States and it is not surprising that some of the first work on the capacity of priority type junctions was carried out in the U.S.A. A measure of the practical capacity of an intersection is the average delay to minor road vehicles. At volumes beyond the practical capacity the average delay increases considerably with only small increases in volume.

Morton S. Raff of the Bureau of Highway Traffic, Yale University and Jack W. Hart[1] of the Eno Foundation for Highway Traffic Control made observations on the delays to vehicles at four intersections in the built-up areas of New Haven, Connecticut. They hoped to obtain an empirical equation connecting minor and major road volumes with average delays to minor road vehicles. Observations were made using a multi-pen recorder to note the time at which vehicles passed through the intersection and it was possible by this means to note traffic volumes and delays.

Typical of their observations is figure 18.1, which shows the average wait for all minor road vehicles. The considerable scatter of the results could not be explained by the investigators and they were, not surprisingly, unable to obtain an empirical relationship connecting main road volume and average delay.

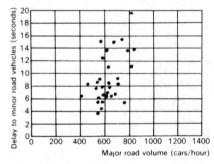

Figure 18.1   Observed variations in delay to minor road vehicles (adapted from ref. 1)

160

An analytical approach to delay

It was not until 1962 that a mathematical formula or model was proposed which connected the many variables inherent in calculating the average delay to side road vehicles. J. C. Tanner[2] proposed a formula for the average delay $\bar{w}_2$ to minor road vehicles.

The variables used to calculate $\bar{w}_2$ in this formula are

The major road flow q_1.
The minor road flow q_2.
The minimum time headway between major road vehicles β_1.
The minimum time headway between minor road vehicles emerging from the minor road β_2.
The average lag or gap α in the major road traffic stream accepted by minor road drivers when entering the major road traffic stream.

These variables and their determination for a particular junction will now be explained.

No explanation of the calculation of q_1 and q_2 is required but it should be noted that the formula for delay was propounded for a T-junction where the merging or crossing movement is into or through a single unidirectional traffic stream. In the formula these flows are normally substituted as vehicles per second.

The introduction of a minimum time headway β_1 between major road traffic introduces an element of bunching or platooning into the flow caused by the inability of vehicles to overtake. Time headway is the interval of time between the fronts of successive vehicles passing a point on the highway. The usual method of estimating β_1 is to note the time headway between the front and rear vehicle of a bunch or platoon of vehicles which are following each other at their minimum separation. The average time headway between them is then calculated. The process is repeated for a considerable number of bunches until an average value representative of the traffic flow is obtained.

The minimum time headway β_2 between minor road vehicles emerging into the major road represents the time between successive minor road drivers who perceive the gap or lag available to them in the major road flow and follow each other into the major road. It is estimated by observing the average time headway between minor road vehicles following each other into the major road.

The last variable needed to estimate the average delay at a priority intersection is the average gap or lag α in the major road stream that is accepted by the minor road driver.

Observations are made of the time at which a minor road vehicle arrives at the junction with the major road and also of the time at which the next major road vehicle arrives at the junction. The gap or lag is the difference between these two times. It is also observed whether the minor road driver enters the first gap or lag that is available to him in the major road traffic stream. If he enters the major road traffic stream, then he is said to have accepted the first available gap or lag, while if he waits until a subsequent gap to enter the major road traffic he is said to have rejected the first available gap or lag. To avoid bias in favour of slower drivers, once a driver has rejected a gap his subsequent performance is neglected.

A lag is an unexpired portion of a gap that is presented to a minor road driver arriving at the junction after the commencement of a gap.

Gaps and lags are classified into classes of suitable size and the percentage of acceptance for each class calculated. When the percentage acceptance is plotted against the gap or lag class mark the curve obtained is of a cumulative normal form and it is possible by statistical means to estimate the most likely value of the 50 percentile acceptance.

With these variables known it is possible to estimate the average delay to minor road vehicles at varying major and minor road traffic volumes using the formula

$$\bar{w}_2 = \frac{\frac{1}{2}E(y^2)/Y + q_2 Y \exp(-\beta_2 q_1)[\exp(\beta_2 q_1) - \beta_2 q_1 - 1]/q_1}{1 - q_2 Y[1 - \exp(-\beta_2 q_1)]}$$

$$E(y) = \frac{\exp[q_1(\alpha - \beta_1)]}{q_1(1 - \beta_1 q_1)} - \frac{1}{q_1}$$

$$E(y^2) = \frac{2 \exp[q_1(\alpha - \beta_1)]}{q_1{}^2(1 - \beta_1 q_1)^2} \{\exp[q_1(\alpha - \beta_1)] - \alpha q_1(1 - \beta_1 q_1) - 1$$

$$+ \beta_1 q_1 - \beta_1{}^2 q_1{}^2 + \tfrac{1}{2}\beta_1{}^2 q_1{}^2/(1 - \beta_1 q_1)\}$$

$$Y = E(y) + 1/q_1$$

In addition Tanner derived an expression for the maximum discharge from a minor road and this is given below, where the symbols have the same meaning as before.

$$q_2(\max) = \frac{q_1(1 - \beta_1 q_1)}{\exp[q_1(\alpha - \beta_1)][1 - \exp(-\beta_2 q_1)]}$$

An indication of likely values of these function flow parameters can be obtained from information that was given in a previous Department of Transport junction capacity assessment method when β_1 was taken as 2 s for minor road vehicles that merge with or cross one major traffic stream and as 1 s when they cross two major road traffic streams. β_2 is taken as 3 s and the value of α varies between 4 s and 12 s according to intersection layout and design speed.

The interactive approach to capacity

Whilst the previously discussed gap acceptance approach describes an important feature of the operation of priority intersections there are a number of difficulties in the use of these methods for estimating the capacity of a junction. Accurate measurement of the parameters that influence capacity are not easy to observe and so there are inaccuracies in the estimation of capacity. Also under heavily-trafficked conditions some major road vehicles give way to the more aggressive minor road drivers and gap acceptance operation breaks down.

An investigation into the relationship between the capacity of non-priority flows and priority flows at major/minor junctions has been carried out by the Transport and Road Research Laboratory and Martin and Voorhees Associates[3]. Subsequently the results of the work were incorporated into a Department of Transport advice note on the design of major/minor junctions[4].

In this approach to design, the capacity of a junction is taken to be the traffic situation when there is a continuous queue feeding one or more turning movements at the junction. Not all the movements need be in this state for the junction to be considered to be at capacity.

Consider the traffic streams shown in figure 18.2 when a major road has arms A and C and a minor road is denoted by arm B. The traffic flows at the junction are represented by q_{c-A} etc. as shown and the streams for which it is required to calculate capacity are

$$q_{b-c}, q_{b-a}, q_{c-b}$$

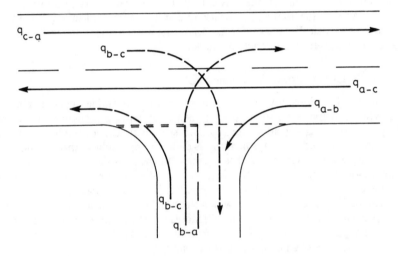

Figure 18.2

In this interactive approach the capacities of these streams depends on the flows of one or more of the priority streams and also on the geometric feature of the junction.

The best predictive equations for turning stream capacity (sat. q) when the major road has approach speeds not exceeding 85 kph are given as:

$$q_{b-a} = D(627 + 14W_{cr} - Y[0.364q_{a-c} + 0.114q_{a-b}$$

$$+ 0.229q_{c-a} + 0.520q_{c-b}]) \tag{18.3}$$

$$q_{b-c} = E(745 - Y[0.364q_{a-c} + 0.144q_{a-b}]) \tag{18.4}$$

$$q_{c-b} = F(745 - 0.364Y[q_{a-c} + q_{a-b}]) \tag{18.5}$$

where $Y = (1 - 0.0345W)$ and W is the width of the major road (in metres).

In the above equations for the saturated capacities the geometric characteristics of the junction are represented by D, E and F and are stream-specific:

$$D = [1 + 0.094(wb - a - 3.65)] \; [1 + 0.0009(Vrb - a - 120)]$$
$$[1 + 0.0006(Vlb - a - 150)]$$

$$E = [1 + 0.094(wb - c - 3.65)] \; [1 + 0.0009(Vrb - c - 120)]$$

$$F = [1 + 0.094(wc - b - 3.65)] \; [1 + 0.0009(Vrc - b - 120)]$$

where $wb - a$ denotes the average lane width over a distance of 20 m available to waiting vehicles in the stream $b - a$, and $Vrb - a$, $Vlb - a$ are the corresponding visibilities to the right and left measured in the normal manner for visibility splays. In these equations capacities and flows are in pcu/hour where a heavy goods vehicle is equivalent to 2 pcu. Detailed information on the measurement of geometric characteristics is given in references 3 and 4.

Queues and delays at oversaturated junctions

In the study of junctions it is frequently important to determine queue lengths and delays. When demand is less than capacity then delay and consequently queue length can be predicted using a steady-state approach such as that theorised by Tanner[2]. Using steady-state theory, delays increase as demand approaches capacity and become infinite when demand reaches capacity. In the real-life situation when demand is close to capacity or even when the capacity is exceeded for short periods, the queue length growth lags behind that predicted by steady-state theory.

Deterministic queueing theory in which the number of vehicles delayed is the difference between capacity and demand ignores the statistical nature of vehicle arrivals and departures and seriously underestimates delay. Using deterministic theory, delay is zero when demand equals capacity.

In many traffic situations demand is close to capacity and even exceeds it for short periods of time and a combination of both steady-state and deterministic theory has been proposed by Kimber and Hollis[5]. Using a coordinate transformation technique they developed the following expressions for queue length and delay:

$$\text{Queue length} = \frac{1}{2} \left((A^2 + B)^{\frac{1}{2}} - A \right) \qquad (18.6)$$

where

$$A = \frac{(1 - \rho)(\mu t)^2 + (1 - L_0)\mu t - 2(1 - C)(L_0 + \rho\mu t)}{\mu t + (1 - C)}$$

$$B = \frac{4(L_0 + \rho\mu t)[\mu t - (1 - C)(L_0 + \rho\mu t)]}{\mu t + (1 - C)}$$

$$\text{Delay per unit time} = \frac{1}{2} \left((F^2 + G)^{\frac{1}{2}} - F \right) \qquad (18.7)$$

where

$$F = \frac{(1 - \rho)(\mu t)^2 - 2(L_0 - 1)\mu t - 4(1 - C)(L_0 + \rho\mu t)}{2(\mu t + 2(1 - C))}$$

$$G = \frac{2(2L_0 + \rho\mu t)[\mu t - (1 - C)(2L_0 + \rho\mu t)]}{\mu t + 2(1 - C)}$$

where $C = 1$ for random arrivals and service,
$\quad\quad C = 0$ for regular arrivals and service,
$\quad\quad \mu$ = capacity,
$\quad\quad \rho = q/\mu$,
$\quad\quad q$ = demand,
$\quad\quad t$ = time.

Demand on both major and minor roads increases during peak periods and then decreases, resulting in a decrease and an increase in capacity.

For any non-priority stream it is possible to calculate capacity by taking into account the levels of flow in the other interacting traffic streams and the geometric characteristics of the junction. If the variation of flow with time is known then the changes in queue length and delay can be calculated.

The computer program PICARDY has been developed by the Traffic Engineering Department of the Transport and Road Research Laboratory to determine capacities, queues and delays at 3-arm major/minor priority junctions. It can cater for a single or dual carriageway major road, with or without a right-turning lane, and with a one or two lane approach on the minor road. The program models the build-up and decay of queues and delays through a peak period which can be examined at discrete time intervals through the period, normally at 5 to 15 minute intervals as selected by the user.

An example of the manual computation of queue length and delay using the results of interactive theory follows. At a junction of the type shown in figure 18.2 the demand flows shown in table 18.1 (in pcu) are noted and the capacity is then calculated from equation 18.4.

TABLE 18.1

Time	Demand b–c	Demand a–c	Capacity b–c
0815	200	900	508
0820	500	1000	481
0825	300	1500	350
0830	300	1400	376
0835	200	1100	455
0840	200	900	508

For each time interval the values of q, μ and ρ are calculated and given in table 18.2 and the queue length and delay calculated from equations 18.6 and 18.7.

TABLE 18.2

Time interval commencing	q_{b-c} (pcu/s)	μ_{b-c} (pcu/s)	ρ_{b-c}	Queue length at end of interval	Delay per unit time (S)
0800	0.056	0.141	0.397	0	1.6
0815	0.139	0.134	1.037	40.5	13.0
0830	0.083	0.097	0.856	34.3	26.4
0845	0.083	0.104	0.798	13.6	11.5
0900	0.056	0.126	0.444	1.70	1.0
0915	0.056	0.141	0.397		0.7

In the calculation of queue length and delay it was assumed that the queue was zero at 0800 hours and that vehicles arrived and departed at random. The calculated values are entered in table 18.2.

In the design of junctions the Department of Transport recommend[4] that a Design Reference Flow be selected to allow the detailed design of alternative feasible junctions to be carried out, the performance of which can then be assessed. When a peak hourly flow is selected to represent the Design Reference Flow the traffic function of the road should be considered. If a junction is designed to be adequate for the highest peak hour flow in a future design year then it is likely that the design will not be economically viable.

If the road has a recreational facility where peak summer flows are very much higher than at other times of the year, then it is suggested that the 200th highest hourly flow might be appropriate. On the other hand on an urban road where there are not likely to be very great variations in flow then the 30th highest flow is suggested as being suitable. The 50th highest flow is considered to be most appropriate for inter-urban roads.

References

1. M. S. Raff and J. W. Hart, *A volume warrant for urban stop signs*, The Eno Foundation for Highway Traffic Control, Saugatuck, Connecticut, U.S.A. (1950)
2. J. C. Tanner, A theoretical analysis of delays at an uncontrolled intersection, *Biometrika*, **49** (1962), 163–70
3. R. M. Kimber and R. D. Coombe, The traffic capacity of major/minor priority junctions, *Transport and Road Research Laboratory Supplementary Report* 582, Crowthorne (1980)
4. Department of Transport, Junctions and Accesses: Determination of Size of Roundabouts and Major/Minor Junctions, *Departmental Advice Note* TA 23/81 (1981)
5. R. M. Kimber and Erica M. Hollis, Traffic queues and delays at road junctions, *Transport and Road Research Laboratory Report* 909, Crowthorne (1979)

Problems

1. State the traffic flow characteristics defined by Tanner in the estimation of delay to non-priority vehicles at a priority junction.
2. Describe the difference between lag and gap acceptance when observations are made at a priority junction.
3. Delays at priority junctions can be estimated by probabilistic or deterministic approaches. Discuss the differences in these methods.

Solutions

1. The five traffic-flow characteristics used to calculate the theoretical delay to minor road vehicles at a priority intersection are

q_1 the major road flow with which minor road vehicles confict,
q_2 the minor road demand flow,
β_1 the mean minimum headway between bunched vehicles on the major road,
β_2 the mean minimum headway between bunched vehicles on the minor road
  entering the major road when the major road flow is zero,
α the mean gap or lag accepted by minor road drivers.

2. A lag is a portion of a gap; when a minor road vehicle arrives at the stop or giveway line, then normally the minor road vehicle will accept or reject a portion of the gap between successive major road vehicles. If the driver rejects the offered lag it is a gap between major road vehicles which the minor road driver must either accept or reject. First decisions of drivers are normally lags whilst subsequent decisions are gaps.
3. A probabilistic approach to determining delay at priority junctions takes account of the random or semi-random nature of vehicle arrivals at the junction. A well known approach using probabilistic theory is that due to Tanner which predicts delays rising rapidly as the minor road demand flow approaches capacity and infinite delays at capacity. In contrast, deterministic approaches predict zero delay and queue length when the minor road demand flow equals capacity with queue length and delay increasing with the passage of time. In practice a compromise between the two approaches is used, the relationship between queue length and delay and demand flow being derived by an axis transformation technique.

19

A simulation approach to delays at priority intersections

Prediction of capacities and delays at highway intersections under priority control may be made by the use of empirical, mathematical or simulation models.

The dependent variable in these models is usually the average delay to minor road vehicles and measures the effectiveness of the junction in allowing conflicting traffic streams to merge or intersect. The performance of the junction may be measured and alternative geometric designs evaluated by consideration of average delays to minor road vehicles.

Empirical models attempt to predict highway capacity on the basis of past observations and probably the best known example of this approach is the Highway Capacity Manual[1] calibrated by traffic data collected throughout the United States. Other researchers[2-5] have studied traffic flow at priority type intersections and used regression analysis methods to obtain relationships between average delay and traffic volumes.

At the present time the average delay to minor road vehicles and the maximum discharge from the minor road at a priority intersection are calculated using mathematical models given by Tanner[6].

Both mathematical and empirical models have limitations in their use. Traffic flow at an intersection is such a complex phenomenon that even the most complex mathematical model has to adopt a macroscopic approach in that all vehicles and drivers generally have the same characteristics.

An empirical model would require a considerable amount of data for its calibration and it is often very difficult to obtain this data with the precision, in the quantity or at the required traffic volume levels. Often the traffic volumes required may, when found, not persist long enough for sufficient observations to be obtained.

If the effect of geometric features is being investigated it will also be necessary for a junction having these features to be located or else constructed in the field or on a test track before traffic observations can be obtained.

While the development and verification of a simulation model is a complex process it is not subject to many of the disadvantages of empirical and mathematical models. Simulation models of intersection traffic flow can be flexible enough to cover a wide range of highway and traffic conditions. Inputs to the model can be specified to any distribution and the form of traffic control and driver characteristics varied. Simulation time may be as long as desired and several figures of merit may be printed out during the simulation process.

A simulation model requires the formation of a model system which represents the real situation at the site being studied. Two forms of simulation have been used in the study of highway traffic flow, analog simulation and digital simulation. In analog simulation an analogy is made between the real world situation and an analog physical system, the components of the analog physical system interacting in the same manner as in the real situation.

In digital simulation the state of the simulated system is stored in digital form and the situation is updated in accordance with stored instructions or rules of the model. Updating of the system may be carried out in two ways, by regular time scanning or by event scanning.

When regular time scanning is used the state of the traffic system is examined at equal scan intervals and all required vehicle movements calculated for the next interval in time. Updating of the traffic situation is carried out for all components of the model. The process is then repeated for the next time scan period.

The length of the time scan period selected will affect the precision of the model. It must be small enough to include all vehicle actions of significance but if the scan interval selected is too small the computer run time will be increased. The time scan interval selected will depend on the computer resources available, the objects of the simulation process and the method of generation of vehicular headways.

When event scanning is used the events that have the greatest importance in the operation of the program are used to build the program. The time at which the next significant event occurs is determined by the program and the traffic situation updated to that event.

To evolve a simulation program, it is usually most convenient to draw up a flow chart showing the steps in the simulation process. Figure 19.1 shows the flow chart for the simulation of the traffic system where a single stream of left-turning minor road vehicles intersect or merge with a single stream of major road vehicles. In this program regular time scanning is used and the figure of merit employed is the average delay to minor road drivers.

An appreciation of digital computer simulation can be obtained by a manual simulation. While this is laborious to carry out for any but a very limited real time period it illustrates the use that can be made of mathematical representations of headway and gap acceptance distributions in simulation.

As an illustration, the traffic flow at the intersection of a single stream of left-turning minor road vehicles and a stream of major road vehicles may be simulated. The major road flow, headway distribution may be assumed to be represented by the double exponential distribution with the following parameters

$$\bar{t}_1 = 2.5 \text{ s}$$

$$\bar{t}_2 = 5.5 \text{ s}$$

$$L = 0.5$$

where $\bar{t}_1, \bar{t}_2$ and L are the mean time headway of free-flowing vehicles, the mean

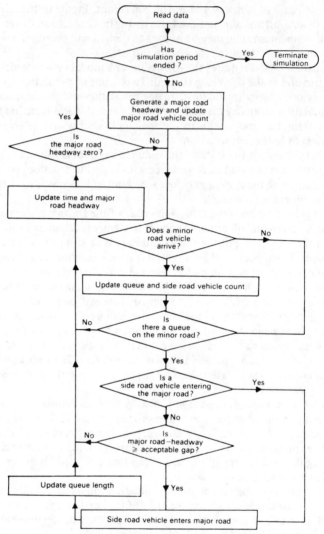

Figure 19.1   A flow chart for the simulation of a single minor road stream with a single major road stream

time headway of restrained vehicles and the proportion of restrained vehicles in the traffic flow respectively.

The minor road flow headway-distribution may be assumed to be represented by the negative exponential distribution, but no vehicle may arrive with a time headway of less than one second between it and the preceding vehicle. The single parameter in this distribution is

$$\bar{t} = 9 \text{ s}$$

where $\bar{t}$ is the mean headway. In addition minor road vehicles exit from the minor road at a minimum headway of 1.0 s.

Gap and lag acceptance for minor road drivers is assumed to be similar and to be represented by the normal distribution with the parameters

$$\bar{t} = 2.5 \text{ s}$$

$$\sigma = 0.9 \text{ s}$$

where $\bar{t}$ and σ are the mean gap and lag acceptance and the standard distribution of gap and lag acceptance respectively.

Furthermore it is noted that no drivers will accept or reject a lag or gap less than the mean minus one standard deviation or greater than the mean plus one standard deviation respectively.

Generation of major and minor stream headways will be carried out by Monte Carlo processes making use of a series of random numbers. Such a series is shown in table 19.1 and could be easily prepared from a series of ten identical cards, each carrying a number from 0 to 9, by picking cards at random and noting the numbers obtained.

TABLE 19.1   A table of random numbers

81	77	72	64	42	31	29	46	62	21
45	93	60	17	35	78	25	42	41	16
11	64	30	58	60	21	33	75	79	74
15	63	47	59	51	13	59	85	27	62
50	53	22	54	96	95	65	24	25	73
27	89	64	72	81	74	40	09	19	61
82	52	75	59	55	79	17	14	24	33
88	66	54	64	70	52	85	50	13	63
23	80	45	68	42	93	67	03	97	42
03	02	59	34	49	77	70	40	75	22
34	95	49	12	91	51	95	80	07	36
51	85	12	18	18	75	06	16	32	84
60	97	67	18	93	92	23	09	41	64
32	78	34	17	83	90	90	51	62	55
58	71	19	68	23	46	76	30	18	79
45	11	28	93	45	58	10	79	31	21
47	56	38	09	90	43	25	27	74	03
52	64	10	66	39	49	93	74	48	15

To generate major road headways, successive two digits are selected from table 19.1 and regarded as the first two decimal places of a random fraction. The random fraction may then be used to solve the following equation for t

$$\text{(random fraction)} = L \exp\left[-(t - e)/(\bar{t}_1 - e)\right] + (1 - L) \exp\left(-t/\bar{t}\right)$$

$$t \geqslant e$$

$$= L + (1 - L) \exp\left(-t/\bar{t}_2\right) \tag{19.1}$$

$$0 < t < e$$

While the rapid solution of this equation to obtain t, the major road headway, would not present any difficulty when using a digital computer a graphical solution will be used in this example.

To obtain the major road headways by graphical means equation 19.1 is plotted as shown in figure 19.2. A series of two random numbers are then taken successively from table 19.1 and inserted in figure 19.2 to obtain the headways which are shown in table 19.2.

Successive headways on the minor road are next simulated by substituting successive random fractions in the cumulative negative exponential distribution

$$(\text{random fraction}) = \exp\left[-(t-1)/(\bar{t}-1)\right]$$

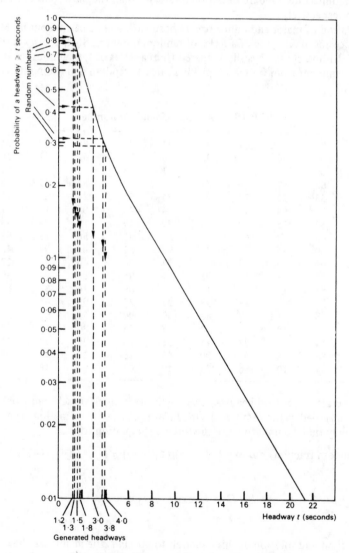

Figure 19.2 The generation of major road headways

TABLE 19.2 Generated major and minor road headways

Random fraction	Major road headway (seconds)	Cumulative major road headway (seconds)	Minor road headway (seconds)	Cumulative minor road headway (seconds)
81	1.2	1.2	2.6	2.6
77	1.3	2.5	3.1	5.7
72	1.5	4.0	3.6	9.3
64	1.8	5.8	4.5	13.8
42	3.0	8.8	7.9	21.7
31	3.8	12.6	10.3	32.0
29	4.0	16.6	10.8	42.8
46	2.7	19.3	7.1	49.9
62	1.9	21.2	4.8	54.7
21	5.3	26.5	13.4	68.1
45	2.8	29.3	7.3	75.9
93	0.7	30.0	1.5	76.9
60	2.0	32.0	5.0	81.9
17	6.2	38.2	15.0	96.1
35	3.5	41.7	9.3	105.4
78	1.3	43.0	3.0	108.4
25	4.6	47.6	12.0	120.4
42	3.0	50.6	7.9	128.3
41	3.1	53.7	8.0	136.3
16	6.6	60.3	15.6	151.9
11	8.5	68.8	18.6	170.5
64	1.8	70.6	4.5	175.0
30	4.0	74.6	10.5	185.5
58	2.1	76.7	5.3	190.8
60	2.0	78.7	5.0	195.8
21	5.3	84.0	13.4	209.2
33	3.7	87.7	9.8	219.0
75	1.4	89.1	3.3	222.3
79	1.2	90.3	2.8	225.1

As with the major road headways, the successive solution of this equation will be made by a graphical method. The cumulative minor road headway distribution is shown in figure 19.3, and successive series of two random numbers are used as random fractions to generate headways as previously. The minor road headways generated in this way are given in table 19.2.

The third set of variables to be simulated are the lag and gap acceptances of minor road drivers. The same technique as has been used previously for generating headways will be used and successive random fractions are entered into the cumulative acceptance distribution as shown in figure 19.4. The acceptable lags obtained in this manner and the successive random fractions used to generate them are given in table 19.3. It should be noted that the generated acceptable lags cannot exceed the mean lag plus two standard deviations; nor be less than the mean lag minus two standard deviations. Maximum and minimum values are thus 3.4 s and 1.6 s, respectively, and it is thus not necessary to amend any of the values generated.

TABLE 19.3 Random fractions and generated lags and gaps

Random fraction	Acceptable gap or lag (seconds)
0.81	3.3
0.77	3.2
0.72	3.0
0.64	2.8
0.42	2.3
0.31	2.0
0.29	1.9
0.46	2.4
0.62	2.8
0.21	1.7
0.45	2.4
0.93	3.8
0.60	2.7

All the variables required for the simulation of traffic flow have now been generated and it is possible to set up a table of events in which successive features of importance are arranged in order of their occurrence. This simulation table is shown as table 19.4 and initially the columns (1) and (2) are introduced from tables 19.2 and 19.3. Column (3) has the same value as column (2) if a queue does not exist. The value in column (3) is the lowest value of the time of arrival or the exit time of the previous vehicle in the queue plus one second.

Column (4), the required lag, is obtained from the previously generated lags in table 19.3 and the values are entered when a vehicle arrives at the stop or give way line.

A time value is entered in column (5) as soon as the gap or lag in the major road stream is equal to or greater than the required lag or gap. It is assumed that a vehicle leaves the minor road instantly but a further vehicle cannot leave the minor road for a further one second.

Column (6) records the queue length while column (7) is the difference between time of arrival of a minor road vehicle and the time of exit. This column is summed and when divided by the number of minor road vehicles entering the intersection gives the average delay to minor road vehicles.

For the traffic situation simulated the value of average delay is 2.4 s: the period of simulation is however extremely short and the simulation should be extended to a minimum period of 600 s. At high traffic flows the effect of assuming that a queue does not exist at the commencement of the simulation period tends to underestimate delays unless the simulation period is sufficiently long.

References

1. Highway Capacity Manual, *Highway Research Board Special Report* 87 (1965)

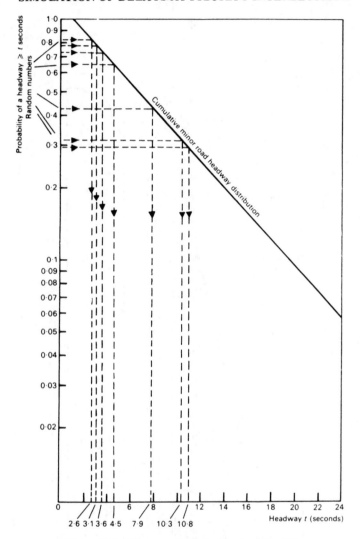

Figure 19.3   The generation of minor road headways

2. W. N. Volk, Effect of type of control on intersection delay, *Proc. Highw. Res. Bd* (1956), 523–33

3. DeLeuw, Cather and Co., The effect of regulatory devices on intersectional capacity and operations, *Final Report Project 3-6*, National Co-operative Highway Research Project, August (1966)

4. J. G. Wardrop and J. T. Duff, Factors affecting road capacity, *International Study Week in Traffic Engineering, Stresa* (1956)

5. F. V. Webster and B. M. Cobbe, Traffic signals, *Tech. Pap. Rd. Res. Bd 56*

6. J. C. Tanner, A theoretical analysis of delays at an uncontrolled intersection, *Biometrika*, **49** (1962), 163–70

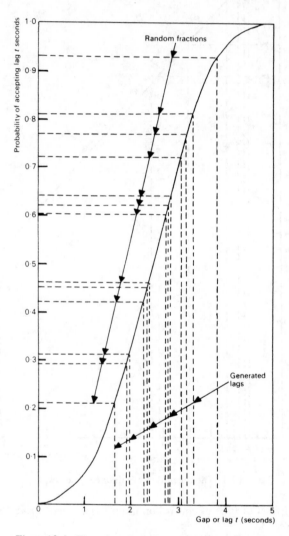

Figure 19.4   The generation of acceptable gaps and lags

TABLE 19.4   The simulation table

1	2	3	4	5	6	7
Major road vehicle arrives at junction	Minor road vehicle arrives at end of queue	Minor road vehicle arrives at stop or give way line	Required gap or lag (seconds)	Minor road vehicle enters major road	Queue length	Delay of existing vehicle (seconds)
1.2						
2.5						
	2.6	2.6	3.3	no	1	
4.0				no		
	5.7			no	2	
5.8				no	2	
8.8				8.8	1	6.2
		9.8	3.2	no	1	
	9.3			no	2	
12.6				12.6	1	6.9
		13.6	3.0	13.6	0	4.3
	13.8	14.6	2.8	no	1	
16.6				no	1	
19.3				no	1	
21.2				21.2	0	7.4
	21.7	22.2	2.3	22.2	0	0.5
26.5						
29.3						
30.0						
32.0						
	32.0	32.0	2.0	32.0	0	0
38.2						
41.7	42.8	42.8	1.9	no	1	

TABLE 19.4 continued

1	2	3	4	5	6	7
Major road vehicle arrives at junction	Minor road vehicle arrives at end of queue	Minor road vehicle arrives at stop or give way line	Required gap or lag (seconds)	Minor road vehicle enters major road	Queue length	Delay of existing vehicle (seconds)
43.0				43.0	0	0.2
47.6						
	49.9	49.9	2.4	no	1	
50.6				50.6	0	0.7
53.7						
	54.7	54.7	2.8	54.7	0	0
60.3						
	68.1	68.1	1.7	no	1	
68.8				68.8	0	0.7
70.6						
74.6						
	75.4	75.4	2.4	no	1	
	76.9			no	2	
76.7				no	2	
78.7				78.7	1	3.3
		79.7	3.8	79.7	0	2.8
	81.9	81.9	2.7	81.9	0	0
84.0						
				cumulative delay		33.0

average delay = cumulative delay/number of minor road vehicles leaving intersection
= 33.0/13 s
= 2.5 s

20

Roundabout intersections

Where traffic flows are small then the control of traffic movements at intersections may be achieved by priority control. As has been discussed previously the form of priority control in this country is that minor road vehicles give way to major road vehicles. On the continent of Europe nearside priority is used where vehicles give way to traffic approaching from the right while in Australia off-side priority is used.

As traffic flows increase delays with priority control become excessive and at high levels of flow grade-separated junctions are necessary. A junction of this form is however extremely expensive: in addition, land requirements are great and in urban and suburban areas interference with pedestrian flow can be considerable. For these reasons at-grade intersections either of the roundabout or signal control type are extremely important in urban areas. The characteristics of these junction types have been described by Millard[1] and a consideration of these characteristics will usually determine which type of junction is appropriate. They include:

(a) In urban situations land requirements are usually the deciding factor. If this is so, it will be found that the land required for a large island roundabout is greater than that needed for traffic signal control. This is especially true if flows on one pair of arms are low. On the other hand, if land purchase is necessary, it is often easier to acquire corner sites necessary for a roundabout than the long narrow strips needed when parallel widening of traffic signal approaches is carried out.

(b) Both conventional and mini-roundabouts have difficulty in dealing with unbalanced flows, especially during peak hours when the traffic entering on an arm is considerably greater than the traffic leaving by it In such situations there is frequently a shortage of gaps in the circulating stream and, under off-side priority rule, delays may become excessive.

(c) Right-turning vehicles (left-hand rule of the road) cause difficulties with signal control when their numbers are large. Either late start or early cut-off facilities or a special phase must be provided causing reduced overall capacity at the junction. In such circumstances roundabouts offer advantages.

(d) Traffic signal control has difficulty in dealing with three-way junctions, especially where the flows are balanced. To a lesser extent this is true of junctions with five or more approaches.

Roundabouts have been commonly used in central city areas where traditionally they were used to resolve traffic and pedestrian conflicts in the large open squares which existed in the early part of the twentieth century. The central island frequently covered a large area of the square and was utilised for ornamental flower beds whilst traffic circulated around the surrounding carriageway. Increasing traffic demand and the pressure to allow pedestrians to cross the carriageway at surface level has resulted in many of these roundabouts being converted to signal control so that more positive control over traffic movements on an area wide basis may be exercised.

In suburban areas, roundabouts are frequently found at the intersection of radial and ring type roads where they are subjected to peak hour traffic demands due to commuter flows.

Roundabouts are also used on rural roads where traffic flows or the road type do not justify the provision of a grade-separated intersection. In these situations, speeds are high on the approaches to roundabouts and safety is an important consideration.

Roundabouts deal efficiently with traffic movements when there are three or four arms With three arms, and well balanced flows a roundabout is considered to be more efficient than signal control. When the number of arms exceeds four then direction signing and driver comprehension become difficult. With many approach arms the diameter of the roundabout increases, leading to possible higher circulating speeds and consequent safety problems

In addition to the resolution of traffic conflicts, roundabouts are employed where there is a significant change in road type, a change from rural to urban conditions or when a significant change in road direction is required.

In current United Kingdom practice there are three basic types of roundabout: normal roundabouts, mini roundabouts and double roundabouts. These are variations of these types to form ring junctions, grade-separated roundabouts and signalised roundabouts.

The Department of Transport[2] defines a normal roundabout as one which has a one-way circulatory carriageway around a kerbed central island 4 m or more in diameter and usually with flared approaches to allow multiple vehicle entry.

Mini roundabouts are defined as a roundabout having a one-way circulatory carriageway around a flush or a slightly raised circular marking less than 4 m in diameter and with or without flared approaches They have been widely used in urban areas where the speed limit does not exceed 30 mph. Physical deflection of vehicle paths to the left, an important factor in roundabout safety, may be difficult in urban junctions with fixed kerb lines and in these circumstances road markings should be used to induce some vehicle deflection. The circular marking varies in diameter from 1 to 4 m diameter and is domed to a maximum height of 125 mm. If space within the junction is very limited then the central island will be frequently run over by larger vehicles and in these cases the island is normally flush with the road surface. The Department of Transport advise that pedal cyclists may experience difficulty and if there are a substantial number of cyclists passing through the junction then signal control may be preferable.

A double roundabout is defined as an individual junction with two normal or mini roundabouts either contiguous or connected by a central link road or kerbed island. It is considered[2] that this form of junction will have advantages in the following circumstances: improving an existing staggered junction where it avoids the need to realign one of the approach roads, unusual junctions such as scissors junctions, joining two parallel routes separated by a river, railway or motorway, existing crossroads where opposing right-turning movements can be separated over-loaded single roundabouts, and junctions with more than four entries.

A solution to the problem of large roundabouts where entries are approaching capacity may sometimes be found by conversion to a ring junction where the usual clockwise circulation of vehicles around a large island is replaced by two-way circulation with three-arm mini roundabouts at the junction of each approach arm with the circulatory carriageway. Because of the two-way flow on the circulatory carriageway and the need for circulating vehicles to give way, adequate signposting is essential for efficient operation.

Roundabout control is also utilised in grade separated intersections in the form of two-bridge roundabouts and dumb-bell roundabouts; these are discussed subsequently in chapter 24.

When roundabouts are being designed it is usual to add at least one additional lane at the entry of the approach roads to the circulating area with a maximum addition of two lanes and a maximum entry width of four lanes. When only low flows are predicted in the future then widening may not be considered necessary but a minimum of two lanes in an entry is desirable The angle at which vehicles enter the circulating area is of considerable importance in the operation of a roundabout. If vehicles enter at an angle approaching 90° to the circulating flow then vehicles are liable to stop quickly on entry and cause rear end collisions. Should they however enter at a small angle then drivers must look over their shoulders in an attempt to merge.

Safety at roundabouts is enhanced by limiting circulating speeds and this is achieved in geometric design by entry path curvature, which is a measure of the vehicle path deflection to the left imposed on vehicles entering the circulating area. The path of vehicles travelling straight through the roundabout is drawn with a flexible curve so that it is the centre line of the most realistic path that a vehicle would take in its complete passage through the junction on a smooth alignment without sharp transitions. The tightest radius of the entry path curvature is recommended by the Department of Transport[2] to be not greater than 100 m.

The size of a roundabout is largely determined by the inscribed circle diameter; the inscribed circle is that which approximately touches the outer edges of the circulating area. For a normal roundabout the smallest recommended inscribed circle diameter for United Kingdom construction is 28 m; if this is not possible a mini roundabout should be used. To meet deflection requirements the inscribed circle diameter should not be less than 40 m. For a mini roundabout the inscribed circle diameter should not be greater than 28 m.

Circulating carriageway widths should be constant and not exceed 15 m; in general the width should not be less than, or 20 per cent greater than the maximum entry width.

It is recommended[2] that the entry radius into the circulating area should lie between 6 m and 100 m with a normal value of approximately 20 m Existing radius should not be less than 20 m and preferably have a value of 40 m.

Accidents at roundabouts

Roundabouts are considered to be one of the safest forms of junctions. The Department of Transport has stated[3] that the average accident cost at a roundabout is approximately 30 per cent less than that at all other junctions and about 60 per cent less than that on the approach carriageways.

In a similar manner to the estimation of personal injury accidents at priority intersections the Department of Transport COBA program estimates personal injury accidents at roundabouts by means of either an inflow (I) or a cross-product (C) model. Both models are of the form

$$\text{Annual number of accidents} = x(f)^y$$

In the inflow model the value of f is the value of the total inflow into the roundabout from all the approach road links expressed as thousands of vehicles AADT. In the cross-product model f is the value obtained by multiplying the combined inflow from two major opposing links by the sum of the inflows from the other links.

The COBA program categorises junctions in a very broad manner and for roundabouts the division is by the number of arms and by the generalised descriptions, standard, small and mini For all three and four-arm roundabouts the formula which is employed is the cross-flow type and the values of the coefficient x and y are given in table 20.1.

A more detailed analysis of accidents at four-arm roundabouts has been carried out by Maycock and Hall[3] who analysed personal injury accident records from a sample of 84 four-arm roundabouts on main roads in the United Kingdom. The roundabouts studied comprised both conventional and small roundabouts in speed limit zones of 30 to 40 and 50 to 70 mph. A cross-product model of the form used

TABLE 20.1   Coefficients for use in accident estimation models

Roundabout type	No. of arms	Highest link standard	x		y	
			rural	urban	rural	urban
Standard	3	single	0.033	0.760	0.033	0.760
	3	dual	0.033	0.760	0.033	0.760
	4	single	0.024	0.890	0.048	0.740
	4	dual	0.063	0.690	0.022	0.850
Small	3	single	0.033	0.760	0.033	0.760
	3	dual	0.033	0.760	0.033	0.760
	4	single	0.101	0.660	0.263	0.540
	4	dual	0.101	0.660	0.263	0.540
Mini	3	single	0.012	1.040	0.012	1.040
	3	dual	0.012	1.040	0.012	1.040
	4	single	0.070	0.640	0.070	0.640
	4	dual	0.070	0.640	0.070	0.640

in the COBA program was evolved and the value of x in the COBA equation was 0.095 for small roundabouts and 0.062 for conventional and dual-carriageway roundabouts; the value of y was found to be 0.68 for all roundabout types. Additionally arm-specific accident relationships for differing vehicle movements were developed and the reader should consult the reference quoted for detailed information.

The traffic capacity of roundabouts

Prior to 1966 traffic entering and passing through a roundabout resolved the vehicle-to-vehicle interactions by requiring the traffic streams to weave one with another as they passed around the central island. In this situation the traffic design of rounda-bouts consisted of designing the individual weaving sections of the carriageway around the central island, an individual weaving section lying between adjacent entry arms.

These weaving sections were designed using a formula developed by Wardrop[4] which incorporated parameters that described the geometric shape of the weaving section and also the proportion of weaving vehicles in the section.

Without any positive control of traffic, roundabouts tended to 'lock' under heavy traffic conditions because vehicles already on the roundabout were prevented from leaving by vehicles attempting to enter. For these traffic conditions a solution was the use of long weaving sections and large roundabouts.

To overcome the problem of locking, the 'give way to traffic approaching from the right' rule was introduced. Subsequently a number of investigations showed that because traffic was no longer weaving around the central island the original formula was no longer valid and the correct approach was to consider the capacity of the roundabout as the capacity of the individual entries of the roundabout.

A particular difficulty in deriving a formula for the capacity of roundabouts is the variety of geometric designs found in practice. Designs range from conventional roundabouts with approximately parallel-sided rectangular-shaped weaving sections, to irregularly shaped central islands and entries on to the roundabout with widths rather similar to that of the approach carriageway, and to the offside priority roundabout with circular central islands and flared approaches.

Recommendations for determining the capacity of a roundabout entry are given by the Department of Transport[5] based on the research carried out by the Transport and Road Research Laboratory[6].

The predictive equation for entry capacity (Q_E) is given by $k (F - f_c Q_c)$ when $F_c Q_c$ is less than or equal to F (otherwise the entry capacity is zero)

where Q_E = the entry flow into the circulatory area in pcu/hour where a heavy
goods vehicle equals two passenger cars,

Q_c = the flow in the circulating area in conflict with the entry in pcu/hour,

k $= 1 - 0.00347 (\Phi - 30) - 0.978 [(1/r) - 0.05]$

F $= 303 X_2$

f_c $= 0.210 t_D (1 + 0.2 X_2)$

t_D $= 1 + 0.5/(1 + M)$

M $= \exp [(D - 60)/10]$

X_2 $= v + (e - v)/(1 + 2S)$

S $= 1.6(e - v)/l^l$

A description of these symbols is given in table 20.1 and a precise definition is given in figure 20.1.

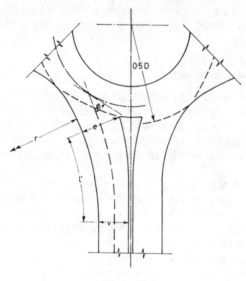

Figure 20.1

The equation for Q_E is applicable to all roundabout types except those that are incorporated into grade-separated intersections. In the latter case the term F is replaced by $1.11F$ and the term f_c is replaced by $1.4f_c$. The formula was derived from observations of traffic flow at roundabout intersections and the range of the geometric parameters at the observed roundabouts are given in table 20.2, together with the range of these parameters suggested for new design. It is also recommended that the circulatory carriageway width around the roundabout should be constant at 1 to 1.2 times the greatest entry width, subject to a maximum of 15 metres.

Whilst the use of the formula for Q_E will allow a check to be made that entry flows are below the entry capacity, there are difficulties when queueing commences. In these circumstances the entry and circulating flows are interdependent.

TABLE 20.2

Symbol	Description	Observed range	Recommended range for design
e	Entry width	3.6–16.5 m	4.0–15.0 m
v	Approach half-width	1.9–12.5 m	2.0–7.3 m
l^I	Average effective flare length	1–∞ m	1.0–100.0 m
S	Sharpness of flare	0–2.9 m	
D	Inscribed circular diameter	13.5–171.6 m	15–100 m
Φ	Entry angle	0–77°	10–60°
r	Entry radius	3.4–∞ m	6.0–100.0 m

To enable delays to be predicted in this interactive situation the Transport and Road Research Laboratory have developed the computer program[7] ARCADY to model queues and delays at roundabouts.

Initially, there is assumed to be no circulating flow past the first entry, so that the entry flow will be either the entry demand or the entry capacity whichever is the least. This entry flow, after removal of vehicles that take the next exit, becomes the circulatory flow past the next entry. The entry flow at this next entry can then be determined by the circulating flow/entry flow relationship. This means that the circulatory flow and hence the entry flow at successive entries can be calculated in a clockwise manner. The process is iterated and the entry flows converge to their final values.

In the program successive short time intervals are considered during which the circulating flow, entry capacity and demand on the entering arm are known; hence the increase or decrease in queue length and the total queue length and delay can be calculated.

Input data for the program is of three types: reference information giving details of the computer run, date and names of roads at the intersection being studied; geometric characteristics of the junction and traffic flow information.

Output for the program includes tables of demand flows, capacities, queue lengths and delays for each time segment, together with a diagrammatic representation of the growth and decay of the queues on each arm.

Example

A roundabout at a grade-separated intersection has an inscribed circle diameter of 65 m, an entry width of 8.5 m and an approach half-width of 7.3 m. The effective length over which flare is developed is 30 m the entry radius is 40 m and the entry angle is 60 degrees.

The design reference flows to be used for a preliminary design expressed as entry capacity (pcu per hour) are given in table 20.3. Calculate the reserve capacity of the intersection in these traffic conditions.

The flows given in table 20.2 are assigned to the roundabout as shown in figure 20.2. From this diagram the circulating flows around the central island are abstracted and these are shown in figure 20.3.

TABLE 20.3

From	To			
	N	S	E	W
N		850	200	100
S	700		450	250
E	150	350		700
W	350	450	350	

The capacity of an entry of a grade-separated roundabout may be calculated using the relationship

$$Q_e = k(F - F_c Q_c)$$

where K $= 1 - 0.00347(\Phi - 30) - 0.978((^1/r - 0.05))$
$= 1 - 0.00347(60 - 30) - 0.978((^1/40) - 0.05)$
$= 1 - 0.1041 + 0.02445$
$= 0.92035.$

S $= 1.6(e - v)/l^1$
$= 1.6(8.5 - 7.3)/30$
$= 0.064.$

$X_2 = v + (e - v)/(1 + 2s)$
$= 7.3 + (8.5 - 7.3)/(1 + 2 \times 0.064)$
$= 8.3638.$

F $= 1.11(303x_2)$
$= 1.11(303 \times 8.3628)$
$= 2813.$

t_D $= 1 + 0.5/(1 + \exp((D - 60)/10))$
$= 1 + 0.5/(1 + \exp((65 - 60)/10))$
$= 1.1888.$

f_c $= 1.4(0.210t_D(1 + 0.2x_2))$
$= 1.4(0.210 \times 1.1888(1 + 0.2 \times 8.3638))$
$= 0.5846.$

$Q_c = 0.92035(2813 - 0.5846Q_c)$
$= 2589 - 0.538Q_c$

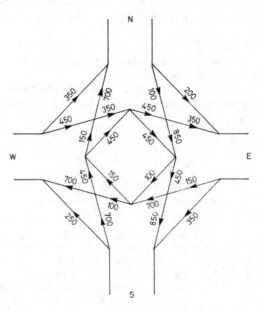

Figure 20.2

Using this relationship between entry capacity and circulating flow the entry capacity for the N, S, E and W arms can be calculated using the values of circulating flow obtained from figure 20.3; calculated values are entered in table 20.4.

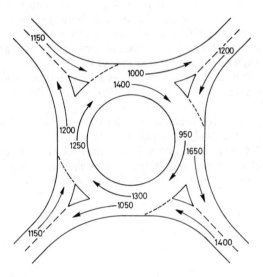

Figure 20.3

TABLE 20.4

Entry	Circulating flow (pcu/h)	Entry capacity (pcu/h)	Entry flow (pcu/h)	Reserve capacity (per cent)
N	1250	1917	1150	66.7
S	950	2078	1400	48.4
E	1400	1836	1200	53.0
W	1300	1890	1150	64.3

Comparison of the calculated entry capacities and the entry flows given in table 20.3 shows that the entry flow is lower than the calculated capacity for each of the entries. The reserve capacity is calculated from capacity minus flow divided by entry flow, expressed as a percentage.

References

1. R. S. Millard, Roundabouts and signals, *Traffic Eng. Control* **13** (1971), 13–15
2. Department of Transport, The Geometric Design of Roundabouts, *Departmental Advice Note* TA 42/84, London (1984)
3. G. Maycock and R. D. Hall, Accidents at 4-arm roundabouts, *Transport and Road Research Laboratory Report* 1120, Crowthorne (1984)
4. J. G. Wardrop, The traffic capacity of weaving sections of roundabouts, *Proceedings of the First International Conference on Operational Research, Oxford, 1957*, English Universities Press, London (1957), pp 266–80
5. Department of Transport, Junctions and Accesses, Determination of size of roundabouts and major/minor junctions, *Departmental Advice Note* TA 23/81, London (1981)
6. R. M. Kimber, The traffic capacity of roundabouts, *Transport and Road Research Laboratory Report* LR942, Crowthorne (1980)
7. E. M. Hollis, M. C. Semments and S. L. Dennis, ARCADY: a computer program to model capacities, queues and delays at roundabouts, *Transport and Road Research Laboratory Report* 940, Crowthorne (1980)

21

Merging, diverging and weaving at grade-separated junctions and interchanges

Merging, diverging and weaving occurs at and between many types of junctions in both urban and rural situations. Most frequently these traffic actions are associated with junctions on routes with a high standard of geometric design such as is found on the British motorway and trunk road system.

In British practice a grade-separated junction is one where the commencement or termination of slip roads connects with an at-grade roundabout or major–minor priority junction. An interchange gives uninterrupted movement from one mainline to another mainline carriageway by link roads using merging and diverging movements.

Weaving between traffic streams can occur in both urban and rural situations and is often inherent in urban traffic management systems.

Design standards for merging, diverging and weaving design in the United Kingdom are given by the Department of Transport[1]. A collective term used for link and slip roads is 'connector roads' and their design speeds are given in table 21.1. In many types of grade-separated junctions and interchanges, vehicles are required to change their direction of motion by travelling around loops, the minimum radius of these loops is related to speed. For loops on to or off a motorway the minimum radius is

TABLE 21.1 Connector road design speed (kph)

	Mainline design speed			
	Urban 100/85		Rural 120/100	
	Link roads	Slip roads	Link roads	Slip roads
Desirable (absolute) minimum speed	70(50)	60(50)	85(70)	70(60)

189

75 m; for loops on to all-purpose roads the value is 30 m, and off all-purpose roads 50 m.

Maximum flow levels used in the design of merging, diverging and weaving areas are based on flows observed in the United Kingdom and differ according to road class as shown in table 21.2.

TABLE 21.2　Maximum values of hourly flow (kph)

Road class	Main line, two lane links and slip roads > 6 m	Single lane links and slip roads < 5 m
All-purpose	1600	1200
Motorway	1800	1350

The proportion of heavy goods vehicles in the flow and uphill gradients affects design and it is necessary to correct design hour flows for these effects using the factors given in table 21.3

TABLE 21.3　Percentage correction factors for gradient

Percentage heavy goods vehicles	Mainline gradient		Merge connector gradient		
	< 2%	> 2%	< 2%	72% < 4%	> 4%
5		+10		+15	+30
10		+15		+20	+35
15		+20	+5	+25	+40
20	+5	+25	+10	+30	+45

The design flow used in merging, diverging and weaving calculations depends upon the road type. For main urban roads as defined by the Traffic Appraisal Manual[2] the 30th highest annual hourly flow is used; for Inter-Urban and Recreational road types the 50th and 200th highest hourly flows respectively are used. These flows should be predicted for 15 years after the opening of the highway.

Design of merging lanes and diverging lanes

In the design of merging and diverging lanes at grade-separated junctions and intersections, the following factors are considered:

(a)　The number of lanes and the design hour flows upstream and downstream of the merging and diverging areas on the main carriageway.
(b)　The number of lanes and the design hour flows on the merging or diverging link or connector.
(c)　The traffic composition and the gradient of the main line carriageway and the merging or diverging links; correction factors are given in table 21.3.

(d) Maximum hourly flows per lane based on United Kingdom experience; these differ for all-purpose roads and for motorways, and details are given in table 21.2.
(e) The relationship between mainline and entry flow for merging lanes and the requirement that merging and diverging flows should not exceed the upstream and downstream flow for merging and diverging lanes respectively.

Using these limiting factors the Department of Transport have developed flow region diagrams for merging and diverging lane design, as shown in figures 21.1 and

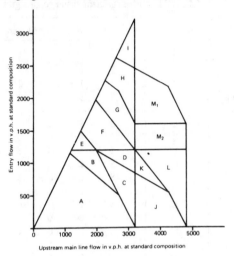

Figure 21.1   Merging diagram (from ref. 3)

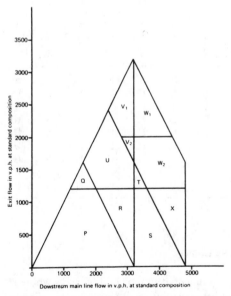

Figure 21.2   Diverging diagram (from ref. 3)

21.2. Because of the differing flow levels for motorways and all-purpose roads, the flow values given in these figures must be multiplied by 1.125 when motorway merges and diverges are being considered.

The flow group indicated from figures 21.1 and 21.2 are inserted into table 21.4 which indicates upstream and downstream mainline carriageway widths, link and slip road widths and the number of entries in the case of merges.

TABLE 21.4   Merge and diverge designs

		Number of lanes										
Upstream main line		2	2	2	2	3	3	3	3	2	2	3
Link		1	1	2	2	1	1	2	2	1	2	1
Entry		1	2	2	2	1	2	2	2	1	2	1
Downstream main line		2	2	2	3	3	3	3	4	3	4	4
	A	S	S	S	S	S	S	S	S	S	S	S
	B	U	S	S	S	S	S	S	S	S	S	S
	C	U	U	U	S	S	S	S	S	S	S	S
	D	U	U	U	S	U	S	S	S	S	S	S
Flow	E	U	U	S	S	U	U	S	S	U	S	U
region	F	U	U	U	S	U	U	S	S	U	S	U
	G	U	U	U	S	U	U	U	S	U	S	U
	H	U	U	U	U	U	U	U	S	U	S	U
	I	U	U	U	U	U	U	U	U	U	S	U
	J	U	U	U	U	S	S	S	S	U	U	S
	K	U	U	U	U	U	S	S	S	U	U	S
	L	U	U	U	U	U	U	U	S	U	U	S
	M	U	U	U	U	U	U	U	S	U	U	U

Upstream main line		2	2	3	3	3	4	3	4	4
Link		1	2	1	2	2	2	1	1	2
Downstream main line		2	2	3	2	3	3	2	3	2
	P	S	S	S	S	S	S	S	S	S
	Q	U	S	U	S	S	S	U	U	S
	R	U	U	S	S	S	S	S	S	S
	S	U	U	S	U	S	S	U	S	U
	T	U	U	U	U	S	S	U	U	U
	U	U	U	U	S	S	S	U	U	S
	V	U	U	U	U	U	S	U	U	S
	W	U	U	U	U	U	S	U	U	U
	X	U	U	U	U	U	S	U	S	U

S = Satisfactory
U = Unsatisfactory

Design of weaving sections

Weaving takes place in many situations on highways where traffic streams merge, diverge and cross whilst travelling in the same general direction. This form of weaving is to be found on main carriageways between intersections and is caused by conflicts in the paths of entering, leaving and straight through vehicles. It is also to be found on link roads within free-flow intersections and in the area of some junctions where entering and leaving vehicles conflict.

To design the weaving-section it is necessary to know the design speed of the mainline carriageway upstream of the weaving area; the mainline is normally taken as the one carrying the major flow. The maximum allowable hourly flows per lane used in the design of weaving sections are given in table 21.2 and vary with carriageway width and road type.

The design of a weaving section makes use of two graphs which are given in reference 1. The length of the weaving section is given by the larger graph of figure 21.3 which relates the minimum length of the weaving section to the total weaving flow for differing ratios of the maximum allowable hourly flow per lane to the upstream design speed. The smaller graph of figure 21.3 relates the length of the weaving section to the design speed, the greater of the two weaving lengths is then used for design. On rural motorways the desirable minimum weaving length is recommended as 2 kilometres but in extreme cases when the predicted traffic flows are at the lower end of the range given for the carriageway width being considered then an absolute minimum length of 1 kilometre may be considered.

To calculate the required width of the weaving section the minor weaving flow is multiplied by a weighting factor which takes into account the reduction in traffic flow caused by weaving. This factor depends upon the ratio of the minimum length to the actual length of the weaving section.

The number of lanes required is given by

$$N = \frac{Q_{nw} + Q_{w1}}{D} + \left(\frac{2 \times L_{min}}{L_{act}} + 1 \right) \frac{Q_{w2}}{D}$$

where N = required number of traffic lanes
Q_{nw} = total non-weaving flow (vph)
Q_{w1} = major weaving flow (vph)
Q_{w2} = minor weaving flow (vph)
D = maximum allowable mainline flow (vph/lane)
L_{min} = minimum weaving length from figure 21.3 (m)
L_{act} = actual weaving length available (m)

It can be seen that the maximum value of the weighting factor is 3 when the actual length of the weaving section is equal to the minimum length as given by figure 21.3. The Department of Transport gives the following advice when the value of N is not an integral[3]. Obviously if the junction can be moved then the actual weaving length will change and the value of N can approximate to a whole number of lanes. If this is not possible then if the size of the fractional part is small and the weaving flow is low then rounding down is possible, conversely a high fractional part and a high weaving flow will favour rounding up to an additional lane. Consideration should be given to the uncertain nature of future traffic predictions, the difficulty of obtaining land in urban areas and the difference between urban commuter and recreational flows.

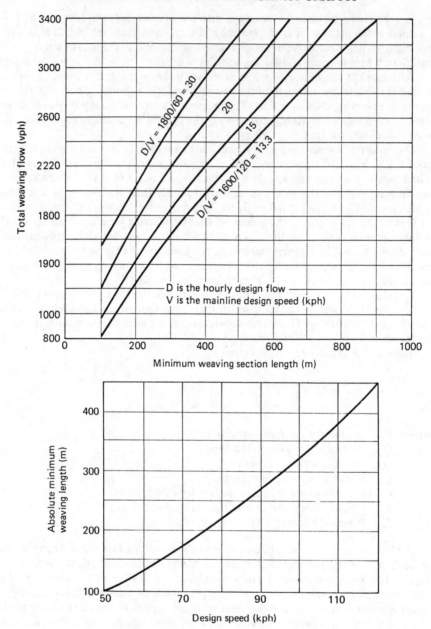

Figure 21.3 Weaving section lengths

References

1. Department of Transport, Layout of grade separated junctions, *Departmental Standard* TD 22/86 London (1986)

2. Department of Transport, *Traffic Appraisal Manual*, London (1982)
3. Department of Transport, Layout of grade separated junctions, *Departmental Advice Note* TA 48/86. London (1986)

Problems

1. The design hour traffic flows at a merge at a rural motorway to motorway interchange are: upstream main line 2500 vph (20 per cent heavy goods vehicles), entry link 700 vph (15 per cent heavy goods vehicles). The two design hour flows coincide in time, the mainline gradient is 3 per cent uphill and the link gradient is 3 per cent downhill. Select a suitable design configuration.

2. The design hour traffic flows through a weaving section on a rural all-purpose link road are: total non-weaving flow 3000 vph, major weaving flow 1500 vph, minor weaving flow 1000 vph. The percentage of heavy goods vehicles in the flow is 10 per cent and the gradient through the weaving section is 1 per cent. The length of the weaving section is 700 m. Determine the required width.

Solutions

1. From table 21.3 the corrected upstream main line flow is

$$2500 \times 1.25 = 3125 \text{ vph}$$

The corrected entry link flow is 700 vph.

Because the interchange is built to motorway design standards these flows must be divided by 1.125 before insertion in figure 21.1, giving 2778 vph and 622 vph. Figure 21.1 indicates these flows fall into flow region A. Table 12.4 indicates a wide range of designs, the simplest of which is 2 lanes on the upstream and downstream main line with a single lane link and entry.

2. From table 21.3 a correction for heavy goods vehicles and for gradient is not required. From figure 21.3 for a total weaving flow of 2500 vph, a main line speed of 120 kph and a maximum value of hourly flow of 1600 vph (tables 21.1 and 21.2), the minimum length of the weaving section is 560 m.

The number of lanes required in the weaving section is given by,

$$N = \frac{Q_{nw} + Q_{w1}}{D} + \left(\frac{2 \times L_{min}}{L_{act}} + 1 \right) \frac{Q_{w2}}{D}$$

$$= \frac{3000 + 1500}{1600} + \left(\frac{2 \times 560}{700} + 1 \right) \frac{1000}{1600}$$

$$= 2.81 + 1.63$$

$$= 4.44$$

A choice has to be made as to whether to round up or round down the required number of lanes. Many factors would be considered in the practical case: land requirements, the nature of the peak flows, that is, commuter or recreational, and the uncertainty of future predictions.

22

Queueing processes in traffic flow

Queueing theory originally developed by A. K. Erlang in 1909 has found widespread application in the problems of highway traffic flow[1].

In any highway traffic situation it is necessary to know: the distribution of vehicle arrivals into the queueing system; whether the source of vehicle arrivals is finite or infinite; whether queue discipline is first come first served, priority or random selection; the number of service stations whereby the vehicle may exit from the system and the distribution of service times for each service station.

A typical system occurs at the entry or exit of a stream of vehicles into or from a parking garage. The vehicles arrive at the parking garage at random, form a queue and enter or leave the garage on a first-come first-served basis. The times required by vehicles to pass through a garage entrance or exit form an approximation to an exponential distribution.

While the input distribution is assumed as random and the service time distribution is exponential in the above example, the application of queueing theory to traffic engineering has been mainly developed around the regular, random and Erlang distributions.

When vehicles arrive at random the numbers of vehicles arriving in successive intervals of time can be represented by the Poisson distribution and an understanding of queueing theory can be obtained from this simple case of Poisson-distributed vehicle arrivals in a single lane, in which vehicles depart with an exponentially distributed service rate.

Consider a traffic queue where $P(n, t + dt)$ is the probability that the queue contains n vehicles ($n > 0$) at time $t + dt$. There are three ways in which the system could have reached this state if it is assumed that dt is so small that only one vehicle could have arrived or departed. These are:

(a) a vehicle did not arrive or depart in time t to $t + dt$;
(b) the queue contained $n - 1$ vehicles at time t and one arrived in time dt;
(c) the queue contained $n + 1$ vehicles at time t and one departed in time dt.

196

Now with Poisson distributed arrivals

$$P(n) = \frac{(qt)^n}{n} \exp(-qt) \tag{22.1}$$

where $P(n)$ is the probability of n vehicles arriving in time t when the mean rate of vehicle arrival is q. From equation 22.1

$$P(0) = \exp(-q\,dt) \tag{22.2}$$

where $P(0)$ is the probability of zero arrivals in time dt and

$$P(1) = (q\,dt)\exp(-q\,dt) \tag{22.3}$$

where $P(1)$ is the probability of one arrival in time dt.
   Expanding equation 22.2

$$P(0) = (1 - q\,dt + q^2\,dt^2/2! \ldots)$$

Expanding equation 22.3

$$P(1) = q\,dt(1 - q\,dt + q^2\,dt^2/2! \ldots)$$

Neglecting second and higher powers of dt

$$P(0) = (1 - q\,dt)$$

and

$$P(1) = q\,dt$$

Similarly the probability of 0 and 1 departures from the queue are

$$P(0) = (1 - Q\,dt)$$

and

$$P(1) = Q\,dt$$

where Q is the mean rate of departure from the queue. Where $n > 0$ the system can reach a state of n vehicles at time $t + dt$ in the following manner

$$P(n, t + dt) = P(n, t) \times P \text{ (a vehicle does not arrive or depart)}$$
$$+ P(n - 1, t) \times P(\text{a vehicle arrives})$$
$$+ P(n + 1, t) \times P(\text{a vehicle departs})$$
$$= P(n, t)\,((1 - q\,dt)(1 - Q\,dt))$$
$$+ P(n - 1, t)\,(q\,dt) + P(n + 1, t)\,(Q\,dt)$$

Ignoring second and higher powers of dt

$$P(n, t + dt) = P(n, t)\,(1 - q\,dt - Q\,dt) + P(n - 1, t)\,(q\,dt) + P(n + 1, t)\,(Q\,dt)$$

$$\frac{P(n, t + dt) - P(n, t)}{dt} = -P(n, t)\,(q + Q) + P(n - 1, t)\,(q) + P(n + 1, t)\,(Q)$$

$$\tag{22.4}$$

In the limit for the steady-state solution the rate of change is zero. Hence

$$0 = -P(n)(q + Q) + P(n - 1)(q) + P(n + 1)(Q)$$

or

$$(1 + q/Q)Pn = (q/Q)P(n - 1) + P(n + 1) \tag{22.5}$$

Similarly when $n = 0$, there are two ways in which the queue can contain n vehicles at time $t + dt$, a change of type (a) or a change of type (c)

$$P(0, t + dt) = P(0, t)(1 - q\,dt) + P(0 + 1, t)(Q\,dt)$$

$$\therefore \frac{P(0, t + dt) - P(0, t)}{dt} = P(0 + 1, t)(Qt) - P(0, t)(qt)$$

As before the steady state of the queue probability of n vehicles in the system is

$$P(1) = P(0)(q/Q)$$

From equation 22.5 when $n = 1$

$$P(2) = (q/Q)^2 P(0) \tag{22.6}$$

Similarly when $n = 2$

$$P(3) = (q/Q)^3 P(0)$$

By induction it can be shown that

$$P(n) = (q/Q)^n P(0) \tag{22.7}$$

When the queue size may be infinite

$$P(0) + P(1) + P(2) + P(3) + \ldots + P(\infty) = 1$$

From equation 22.7

$$P(0) + (q/Q)P(0) + (q/Q)^2 P(0) + (q/Q)^3 P(0) + \ldots + (q/Q)^\infty P(0) = 1$$

$$P(0)(1/(1 - q/Q)) = 1$$

$$P(0) = (1 - q/Q)$$

Also

$$P(n) = (q/Q)^n (1 - q/Q) \tag{22.8}$$

The expected number in the queue $E(n)$ is given by

$$E(n) = \sum_{n=0}^{\infty} nP(n)$$

$$= 0 \times P(0) + 1 \times P(1) + 2 \times P(2) + \ldots + nP(n)$$

$$= (q/Q)P(0) + 2(q/Q)^2 P(0) + \ldots + n(q/Q)^n P(0)$$

$$= (q/Q)P(0)(1 + 2(q/Q) + \ldots + n(q/Q)^{n-1})$$

$$= (q/Q)P(0)(1/(1 - q/Q)^2)$$

$$= q/(Q - q) \tag{22.9}$$

Because there is a probability that the queue will be zero the mean queue length $E(m)$ will not exactly be one less than the mean number in the queue $E(n)$

$$E(m) = \sum_{n=1}^{\infty} (n - 1)P(n)$$

$$= \sum_{n=0}^{\infty} P(n)n - P(n) + P(0)$$

$$= E(n) - q/Q \tag{22.10}$$

As well as the expected number in the queue and the mean queue length, the waiting time w before being taken into service and the total time in the queue v are of considerable importance in the study of traffic phenomena.

The waiting time distribution may be considered in two parts.

Firstly there is the probability that the waiting time will be zero, which means that a queue does not exist, and

$$P(0) = 1 - q/Q, \qquad n = 0$$

Secondly there is the probability that the waiting time for a vehicle is between time w and time $w + dw$

$$P(w < \text{wait} < w + dw) = f(w)\,dw \qquad n > 0 \tag{22.11}$$

Such a delay is possible as long as there is a vehicle in service, which may be expressed as

$$P(n \geqslant 1) = \sum_{n=1}^{\infty} Pn \tag{22.12}$$

For the waiting time for a vehicle to be exactly between w and $w + dw$ all the vehicles in the queue ahead of the one being considered except the one immediately ahead, must depart in time w and the one immediately ahead must be served in time dw. This is the product of the two probabilities.

$$P(n - 1, w) = \frac{(Qw)^{n-1}}{(n - 1)!} \exp(-Qw)$$

from equation 22.1 and

$$P(1, dw) = Q\,dw$$

Substituting these probabilities in equation 22.11 and summing over equation 22.12

$$f(w)\,dw = \sum_{n=1}^{\infty} P(n)P(n - 1, w)P(1, dw)$$

$$= \sum_{n=1}^{\infty} (q/Q)^n(1 - q/Q) \times \frac{Q\,dw}{(n - 1)!} (Qw)^{n-1} \exp(-Qw)$$

$$= q(1 - q/Q) \, dw \, \exp(-Qw) \sum_{n=1}^{\infty} \frac{(wq)^{n-1}}{(n-1)!}$$

$$f(w) = (q/Q)(Q - q) \exp(-w(Q - q)) \qquad w > 0 \tag{22.13}$$

The moment generating function for waiting times is given by

$$M_w(\theta) = \int_0^{\infty} \exp(\theta w) f(w) \, dw$$

$$= \int_0^{\infty} \exp(\theta w)(q/Q)(Q - q) \exp(-w(Q - q)) \, dw$$

$$= (q/Q)(Q - q) \int_0^{\infty} \exp(-w(Q - q - \theta)) \, dw$$

$$= (q/Q)(Q - q)/(Q - q - \theta)$$

$$M'w(0) = E(w) = q/[Q(Q - q)] \tag{22.14}$$

The average time an arrival spends in the queue is given by $E(w)$ plus the average service time $1/Q$

$$E(v) = 1/(Q - q) \tag{22.15}$$

For more generalised cases when the service time can no longer be described by a negative exponential distribution the expected number in the queue when arrivals are random is given by

$$E(n) = \frac{q}{Q} + \left(\frac{q}{Q}\right)^2 (1 + C^2) / \ 2\left[\left(1 - \frac{q}{Q}\right)\right] \tag{22.16}$$

where C is the coefficient of variation of the service time distribution, that is the ratio of the standard deviation to the mean.

If the service is exponential then $C^2 = 1$ and equation 22.16 reduces to equation 22.9.

If the service is regular $C^2 = 0$ and

$$E(n) = \frac{q}{Q} \left(1 - \frac{q}{2Q}\right) / \left(1 - \frac{q}{Q}\right) \tag{22.17}$$

In this case it has been shown that the average time a vehicle spends queueing is given by

$$E(w) = q/2Q(Q - q) \tag{22.18}$$

To illustrate the use of queueing theory in highway traffic flow it is necessary to find some situation in which vehicles are delayed and allowed to proceed in accordance with the simple situations previously considered.

Typical situations occur when vehicles have to stop to enter or leave parking facilities during a period of exceptional demand. Queues can frequently be observed at the entrance to or exit from car parks on public holidays, at weekends at the coast and at sporting events.

It is desirable that the rate of arrival q is reasonably constant and, to make this possible, observations may be divided into shorter periods of time, each with separate value of q.

In some cases the service rate may vary with demand but where queues are forming it is usually possible to assume that the service rate is approximately constant.

As an example of the application of queueing theory observations were made of the number of vehicles waiting to enter two parking areas. One was a multistorey parking garage equipped with automatic entry control equipment and the other was a surface car park where drivers paid the attendant as they entered. In both instances vehicles were delayed as they queued to enter the park and drivers on the approach had considered only one choice of entry gate.

Observations made of vehicle arrivals at the entrance to the surface car park are included to demonstrate the statistical technique whereby vehicle arrivals are shown to be random or non-random.

Where vehicles arrive at random then the numbers of vehicles arriving in successive time intervals may be represented by the Poisson distribution.

Then the probability of n vehicles arriving in a given interval of time t may be calculated from

$$P(n) = \frac{(qt)^n \exp(-qt)}{n!} \tag{22.19}$$

This distribution is often referred to as the counting distribution because it describes the number of vehicles arriving at a given point on the highway.

The numbers of vehicles arriving at the entrance to the surface car park or at the end of the queue in successive 60-second intervals were observed. At this situation marked changes in the arrival rate were not expected and so observations were continued for a period of 3000 seconds and the mean arrival rate taken as q.

To test the form of the arrival distribution and also to determine the mean rate of arrival the observations were tabulated as shown in table 22.1.

Two further sets of data were obtained at this site and the values of both q and Q derived from the observations are included in table 22.4.

It can be seen from tables of chi-squared that there is no significant difference at the 95 per cent level between the observed and the theoretical distributions.

The mean time taken for a vehicle to enter the car park, the service time, was observed as the time interval between successive vehicles in the queue moving away from the attendant or passing beneath the raised barrier in the case of the parking garage.

Table 22.2 gives the observed distribution of service times at the entrance to the surface car park. The general exponential nature of the service time can be seen. Over 50 per cent of drivers are able to pay the attendant and receive a receipt in less than 8 seconds. A smaller number of drivers require a longer period to tender the fee and receive change. No driver takes longer than 28 seconds to enter the car park after arriving at the entrance.

TABLE 22.1 Vehicle arrival distribution at the entrance to a surface car park

No. of vehicles arriving in a 60 s interval	Frequency of observed intervals		Theoretical frequency $(P(n)\,\Sigma f_0)$	Chi-squared
(n)	(f_0)	$(f_0 n)$	(f_t)	(χ^2)
0	1 ⎫	0	0 ⎫	
1	2 ⎬	2	2.40 ⎬	0.01
2	4 ⎭	8	4.86 ⎭	
3	9	27	7.71	0.22
4	10	40	9.18	0.07
5	7	35	8.74	0.35
6	6	36	6.94	0.13
7	4 ⎫	28	4.72 ⎫	
8	3 ⎪	24	3.83 ⎪	
9	2 ⎬	18	1.49 ⎬	0.01
10	1 ⎪	10	0.71 ⎪	
11	1 ⎭	11	0 ⎭	
	$\Sigma\ 50$	$\Sigma\ 239$		$\Sigma\ 0.79$

$$\text{mean headway} = \frac{60 \times 50}{239} = 12.6 \text{ s}$$

$$q = 1/12.6 = 0.08 \text{ veh/s}$$

$$\text{arrival volume} = 3600/12.6 = 286 \text{ veh/h}$$

Where the service time distribution is exponential the probability of drivers requiring service times between the class limits $t - \Delta t/2$ and $t + \Delta t/2$ may be calculated from

$$\frac{\Delta t}{t_s} \exp - \frac{t}{t_s}$$

where Δt is the class interval,
$\quad\quad$ t is the class mark,
$\quad\quad$ $\bar{t}_s$ is the mean service time.

Observed and theoretical values are compared in table 22.2 and it can be seen from tables of χ^2 that there is no significant difference at the 5 per cent level.

Observations of the service time at the entrance to the multistorey parking garage show a considerably different form of distribution. In this case the drivers had only to drive into the entrance bay, take a ticket and drive into the garage when the barrier had been automatically raised. Table 22.3 gives details of these observed service times for one of the 3000-second periods of observation made at this garage.

At the same time as observations were made of the arrival and service time distributions a note was made of the queue length at 100-second intervals.

The average delay to vehicles entering the car park was calculated from

$$\frac{\Sigma \text{ sum of queue lengths} \times 100}{\text{no. of vehicles arriving}}$$

TABLE 22.2   Vehicle service distribution at entrance to surface car park

Service time class interval (seconds)	Observed frequency (f_0)	$f_0 t$	Theoretical frequency f_t	x^2
0–3.9	80	160	87	0.56
4–7.9	53	318	55	0.07
8–11.9	41	410	35	1.03
12–15.9	28	392	22	1.64
16–19.9	18	324	14	0.29
20–23.9	12	264	9	1.00
24–27.9	8 ⎱	208	6 ⎱	4.00
28	0 ⎰	0	10 ⎰	
	Σ 240	Σ 2076		Σ 8.59

$$\bar{t}_s = 2076/240 = 8.7 \text{ seconds}$$

TABLE 22.3   Vehicle service distribution at entrance to multistorey car park

Service time class interval (seconds)	Observed frequency f_0	$f_0 t$
0–3.9	27	54
4–7.9	120	720
8–11.9	31	310
12–15.9	5	70
	Σ 183	Σ 1154

$$\bar{t}_s = 1154/183 = 6.3 \text{ seconds}$$

The observed delay to queueing vehicles for each 3000-second period of observation together with the mean arrival and service rates are given in table 22.4.

Theoretical delays were calculated assuming an exponential distribution of service times (equation 22.14) and a regular distribution of service times (equation 22.18).

It can be seen from table 22.4 that observed delays at the entrance to the surface car park can be approximately represented by equation 22.14. On the other hand delays at the entrance to the multistorey car park, where service is approximately regular because it is only necessary to take a ticket and move through a barrier, exhibit characteristics midway between those given by equations 22.14 and 22.18.

Using the equations derived from queueing theory it would be possible to form an estimate of delays of higher arrival volumes. This allows a balance to be obtained between delays to queueing vehicles and the cost of opening additional entrances.

TABLE 22.4

Arrival volume (veh/h)	q (veh/s)	Mean service time	Q (veh/s)	Observed average delay	Theoretical delay Eq. 22.14	Eq. 22.18
Surface car park						
288	0.0800	8.7	0.1149	25	20	10
293	0.0814	8.7	0.1149	23	21	11
327	0.0908	8.7	0.1149	35	32	16
Multistorey car park						
506	0.1406	6.3	0.1587	41	55	24
497	0.1381	6.3	0.1587	33	49	21
533	0.1481	6.3	0.1587	61	94	44

Suggested reading

1. P. M. Morse, *Queues, Inventories and Maintenance*, Wiley, New York (1963)
2. D. R. Cox and W. L. Smith, *Queues*, Methuen, London (1961)
3. W. D. Ashton, *The Theory of Road Traffic Flow*, Methuen, London (1966)

Problems

1. A census point is set up on a highway where vehicle arrivals may be assumed to be random and the one-way traffic volume is 720 veh/h. All vehicles are required to stop at the census point while a tag is attached, the operation taking a uniform time interval of 4 s.

(a) Is the expected number of vehicles waiting at the census point 1.4, 2.4 or 3.0?

(b) Is the average waiting time of a vehicle at the census point 6.5 s, 8.0 s or 9.4 s?

(c) The enumerators are replaced by untrained staff and the time taken to attach a tag now has the same mean value but the standard deviation of the time taken is noted to be 2 s. Will the number of vehicles waiting at the checkpoint be increased by more than 50 per cent?

Solutions

1. (a) At a queueing situation in which arrivals are randomly distributed and where the service times are uniform then the expected number of vehicles waiting at the census point is given by equation 22.17

$$E(n) = \frac{q}{Q} \left(1 - \frac{q}{2Q}\right) \Big/ \left(1 - \frac{q}{Q}\right)$$

where q is the mean rate of arrival of vehicles,
     Q is the mean rate of departure of vehicles.

In this example

$q = 720/3600$ veh/s
   $= 0.2$ veh/s
$Q = 1/4$ veh/s
   $= 0\,25$ veh/s

$$E(n) = \frac{0.2}{0.25} \left(1 - \frac{0.2}{2 \times 0.25}\right) \Big/ \left(1 - \frac{0.2}{0.25}\right)$$

$$= 0.8(1 - 0.4)/(1 - 0.8)$$

$$= 2.4$$

23

Geometric delay at non-signalised intersections

Delay to vehicles at intersections is an important factor in the choice of intersection type and frequently attention is focussed on queueing delay which occurs at peak flow times. There is however a further source of delay which is found regardless of traffic conflicts, caused by vehicles slowing down and subsequently accelerating as they negotiate the junction. As this form of delay is determined by the size and shape of the junction, it is referred to as 'geometric delay'. Because geometric delay takes place throughout the whole of the day it often is a substantial portion of the whole of the delay.

In the United Kingdom methods for estimating geometric delay have been largely based on observations made by the Department of Civil Engineering at the University of Southampton. A series of reports on geometric delay at priority junctions, roundabouts and large intersections has been drawn together in Transport and Road Research Laboratory Report 67[1]. In this report two basic methods of analysis have been used. Firstly, category or regression analysis was used to develop predictive equations for the delay for each vehicle manoeuvre in terms of junction geometry and approach link speeds. Secondly, a model was developed for the delays for each element of the manoeuvre and these elements were then summed to give the total journey time for the whole manoeuvre. The delay is then obtained by subtracting the theoretical journey time assuming that the links meet at the centre of the junction.

The category analysis indicated that for light vehicles delays at priority junctions, grade-separated roundabouts and diamond intersections, geometric delay could be estimated from table 23.1.

Regression analysis for at-grade roundabouts, grade-separated roundabouts, trumpet intersections and motorway links gave the following expressions for geometric delay (seconds).

At grade roundabouts: $0.11ED + 0.72(Y - V) + 3.06$
Grade-separated roundabouts: $0.08ED + 1.27(Y - V) + 2.38$
Trumpet intersections: $0.17R + 34$
Motorway links: $0.045ED + 0.4$

TABLE 23.1 Geometric delay (seconds) determined by category analysis

	Left turn		Right turn		Straight ahead	
	side	main	side	main	side	main
Priority junction[a]	7.8	5.7	10.6	6.5	12.2	0
Grade separated roundabout	10[b]		28[b]		11	0
Diamond intersection	15		19		0	0

[a]For all movements except main ahead, add 2 s if mean link speeds > 65 kph, and 1.4 s if visibility is sub-standard.
[b]Add 3 s for a flyover when travelling from the minor road to the motorway.

where ED is the extra distance due to negotiating the intersection (m)
       Y is the average of approach and exit link speeds (m/s)
       V is the average speed within the intersection (m/s)
       R is the loop radius (m)

The synthetic model is based on a set of empirical approximations where the actual speed profiles are represented in simplified forms. For example, for an at-grade roundabout the link approach speed V is assumed constant up to the point at which constant deceleration starts until the vehicle speed reaches the speed V_b at which it enters the circulating section of the roundabout. During its passage around the circulating section a vehicle has an average speed V_{bc} and an exit speed from the circulating section of V_c, whilst in large roundabouts the speed within the circulating section may be higher than at entry and exit. This refinement did not increase the overall accuracy of the model and V_b has been taken as equal to V_c. The vehicle then accelerates at a constant rate until it reaches the link exit speed V_d.

Similar models were adopted for use at trumpet intersections where it was noted that speeds on the loop section were reasonably constant, at priority intersections and at diamond intersections where the points of entry and exit were considered to be coincident. For gyratory systems, motorway links and other types of intersection with complex internal sections, speeds within the intersection were related to geometric features. These features were kerb entry and exit radii, sight distances, inscribed circle diameter, turning radii, entry and exit angles, and lengths of slip roads.

In the observations made to develop the geometric delay models, vehicles were divided into light and heavy vehicles. Light vehicles were defined as those with four tyres or less, and heavy vehicles were other vehicles. For this reason heavy vehicles covered a wide range of vehicle types from six wheel vans to multi-axled articulated and other commercial vehicles including buses and coaches. As a result geometric delay for heavy vehicles varied considerably according to intersection type. In

Transport and Road Research Supplementary Report 810[1] it is recommended that for free-flowing motorway links the delay to a heavy vehicle should be the same as for a light vehicle; for priority junctions and diamond intersections light vehicle delay should be increased by 25 per cent and at all other intersections light vehicle delay should be increased by 15 per cent.

Reference

1. M. McDonald, N. B. Hounsell and R. M. Kimber, Geometric delay at non-signalised intersections, *Transport and Road Research Laboratory Supplementary Report* 810, Crowthorne (1984)

Problem

Calculate the geometric delay at a priority junction for vehicles turning left from the major road to the minor where the following speed and geometric conditions exist.

Speed on major road entry link 20 m/s (V_a)
Speed on minor road exit link 18 m/s (V_d)
Entry kerb radius 10 m (ER)

Solution

Overall geometric delay $= JT - [D_{ab}/V_a + D_{cd}/V_d]$. JT is the overall journey time and is equal to $t_{ab} + t_{cd}$ where a is the point on the major road where vehicles commence to decelerate, b and c are coincident for a priority intersection and represent the junction of the major and minor roads, and d is the point on the minor road where vehicles complete their acceleration and attain the minor road link speed. t_{ab} is the deceleration time and t_{cd} is the acceleration time. D_{ab} is the deceleration distance and D_{cd} is the acceleration distance, and

$$t_{ab} = (V_a - V_b)/[1.06(V_a - V_b)/V_a + 0.23]$$

$$t_{cd} = (V_d - V_c)/[1.11(V_d - V_c)/V_d + 0.02]$$

$$D_{ab} = (V_a^2 - V_b^2)/2[1.06(V_a - V_b)/V_a + 0.23]$$

$$D_{cd} = (V_d^2 - V_c^2)/2[1.11(V_d - V_c)/V_d + 0.02]$$

$$V_b = V_c = 1.67\sqrt{(ER)}$$

$$V_b = V_c = 1.67\sqrt{(10)} = 5.28 \text{ m/s}$$

$$t_{ab} = (20 - 5.28)/[1.06(20 - 5.28)/20 + 0.23]$$

$$= 14.57 \text{ s}$$

$$t_{cd} = (18 - 5.28)/[1.11(18 - 5.28)/18 + 0.02]$$

$$= 15.81 \text{ s}$$

$$D_{ab} = (20^2 - 5.28^2)/2[1.06(20 - 5.28)/20 + 0.23]$$

$$= 184.19 \text{ m}$$

$$D_{cd} = (18^2 - 5.28^2)/2 [1.11(18 - 5.28)/18 + 0.02]$$

$$= 184.06 \text{ m}$$

$$JT = t_{ab} + t_{cd}$$

$$= 14.57 + 15.81$$

$$= 30.38 \text{ s}$$

Overall geometric delay $= JT - [D_{ab}/V_a + D_{cd}/V_d]$

$$= 30.38 - [184.19/20 + 184.06/18]$$

$$= 10.94 \text{ s}$$

24

Grade-separated junctions and interchanges

Grade-separated junctions have been defined[1] as ones which require use of an at-grade junction at the commencement or termination of the slip roads. This at-grade junction, in the form of either a major–minor priority junction or a roundabout together with the slip roads, can produce a diamond junction, a half-cloverleaf junction or a roundabout junction.

In the United Kingdom, decisions as to junction type are no longer made by the use of design standards which relate choice to traffic forecasts. Decisions are made on the basis of relative economic and environmental advantages of competing junction types. The choice of junction choice is particularly difficult because of the sensitivity of delay to the future traffic forecast. Below a flow in the region of 85 per cent capacity, average vehicle delay is relatively uniform whilst above 85 per cent of capacity, average delays increase rapidly. As a result small differences in estimates of future demand flows or junction capacity can have a considerable effect on delay estimates. With grade-separated junctions in urban areas, environmental factors may have considerable influence on the choice of junction type.

Three-way junctions

At a three-way junction the usual layout adopted is the trumpet, illustrated in figure 24.1. It allows a full range of turning movements but suffers from the disadvantage that there is a speed limitation on the minor road right-turning flow due to radius size. Where topographical conditions make it necessary the junction can be designed to the opposite hand, as shown in figure 24.2, but this layout has the disadvantage that vehicles leaving the major road have to turn through a small radius.

Where a limited range of turning movements is required then one of the junctions of the form shown in figure 24.3 may be used. The layout shown in (c) may be used in a situation in which the two carriageways are a considerable distance apart and where a right turn exit on the right-hand lane would not be appropriate.

It may sometimes in the future become necessary to convert a three-way junction to a four-way junction, and the design should allow for this future conversion. A suitable form of junction is the partial bridged rotary shown in figure 24.4. Alterna-

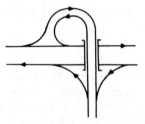

Figure 24.1 The usual form of trumpet intersection

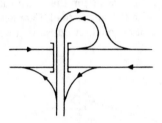

Figure 24.2 A form of trumpet intersection to avoid existing development

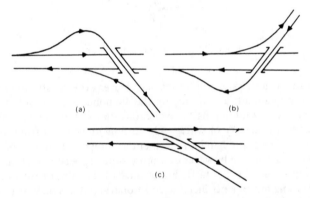

Figure 24.3 Alternative forms of three-way junctions

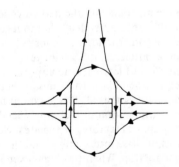

Figure 24.4 A partial bridged rotary intersection

tively the trumpet intersection shown in figure 24.1 may be converted to a cloverleaf intersection.

Four-way junctions

Frequently four-way junctions are formed by the intersection of a major and a minor road. In such instances it is often possible to allow traffic conflicts to take place on the minor road. The simplest type of intersection where this occurs is the diamond, shown in figure 24.5. This is a suitable form of layout in which there are relatively few turning movements from the major road on to the minor road since the capacity of the two exit slip roads from the major road is limited by the capacity of the priority intersection with the minor road. Care needs to be taken with this and other designs by the use of channelisation, so that wrong-way movements on the slip roads are prevented.

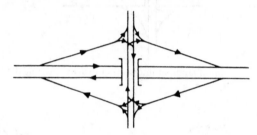

Figure 24.5   A diamond intersection

In some circumstances site conditions may prevent the construction of slip roads in all of the four quadrants as required by the diamond. A solution in this case would be the construction of a half cloverleaf as shown in figure 24.6(a). Where site conditions make it necessary or where right-turning movements from the major road are heavy then the opposite hand arrangement as shown in figure 24.6(b) may be used, with the disadvantage however of a more sudden speed reduction for traffic leaving the major road. As the traffic importance of the minor road and the magnitude of the turning movements increase additional slip roads may be inserted into the partial cloverleaf design until the junction layout approaches the full cloverleaf design.

An advantage of both the diamond and the half-cloverleaf junctions is that only one bridge is required, but it is recommended[1] that in its design provision should be made for the future likely construction of a ghost island on the minor road.

Where major traffic routes intersect, it is no longer possible for the traffic conflicts to be resolved on the road of lesser traffic importance and the simpler junction types previously described are then likely to be unsatisfactory, both from the point of view of capacity and accident potential. The design of intersections of this type is always carried out after consideration of the individual directional traffic flows, having regard for the topographical features of the area. Nevertheless it is possible to discuss the general forms of these intersections and the traffic flows that are likely to warrant their construction. When turning movements from the route of greater traffic importance cannot be handled by the diamond intersection, then the

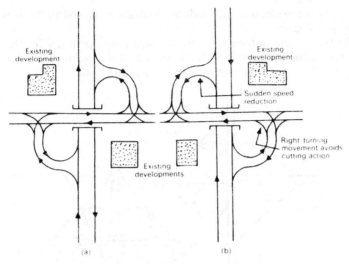

Figure 24.6   Partial cloverleaf designs

usual solution in Great Britain has been the grade-separated roundabout as shown in figure 24.7.

The advantages of this type of junction are that it occupies a relatively small area of land and has less carriageway area than other junctions of this type. It also allows easy U-turns to be made, a factor of some importance in rural areas where intersections may be widely spaced or in urban areas where a number of slip roads may join the major route between interchanges.

As the traffic importance of the minor road increases then a disadvantage of this form of junction is the necessity for all vehicles on the road of lesser traffic importance to weave with the turning traffic from the other route. For this reason the capacity of the weaving sections limits the capacity of the intersection as a whole.

A junction intermediate between the diamond and the two-bridge roundabout is the dumb-bell roundabout where priority junction slip road terminations of the diamond junction are replaced by two compact roundabouts. This form of junction

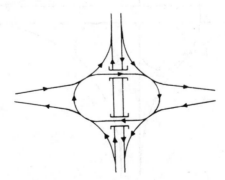

Figure 24.7   Grade-separated roundabout intersection

requires only one bridge and has a reduced landtake compared to the diamond or two-level roundabout.

As the traffic flow on the minor crossing road increases then a disadvantage of the two-level roundabout junction is the necessity for all vehicles on the crossing road to pass through the circulating area of the roundabout.

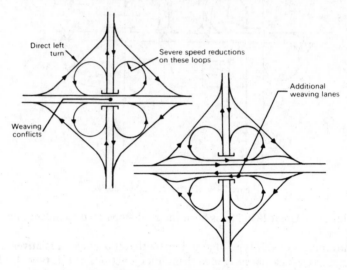

Figure 24.8　Cloverleaf intersections

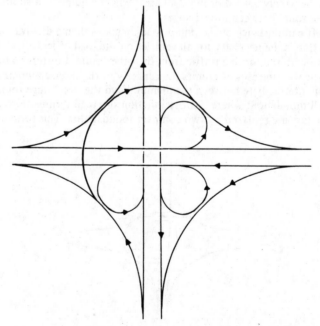

Figure 24.9　A cloverleaf intersection with a direct link for a heavy right-turning movement

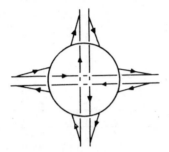

Figure 24.10   A three-level grade-separated intersection

An alternative form of intersection which, while popular in the United States, has found limited application in Great Britain is the cloverleaf, shown in figure 24.8. With this form the straight ahead traffic on both routes is unimpeded and in addition left-turning movements may be made directly from one route to the other. The cost of structural works is also less than with the grade separated roundabout because only one bridge is required although the bridge will normally be wider. On the other hand the carriageway area is greater for the cloverleaf.

Operationally the cloverleaf intersection has many disadvantages. Vehicles leaving both routes to make a right turn have to reduce speed considerably because of radius restrictions and there is also a weaving conflict between vehicles entering and leaving each route, which peak at the same time. This conflict can however be reduced by the provision of a separate weaving lane on either side of each carriageway but this will increase carriageway and bridge costs. There are also four traffic connections on to each route and U-turns may present some difficulty to drivers unfamiliar with the junction. Sign posting has also been stated to present some difficulties.

Where heavy right-turning movements in one direction are anticipated at a cloverleaf type intersection this movement has frequently been given a direct connection by a single link as shown in figure 24.9, requiring extra bridgeworks but eliminating one loop with its small radius and restricted speed.

Where two major routes of equal importance intersect, a solution which has been frequently adopted is the use of a roundabout to deal with turning movements while straight flows on both major routes are unimpeded. This layout is shown in figure 24.10.

The three-level roundabout has advantages compared with larger and more complex interchanges because of its lower carriageway area and reduced landtake. It has however high structure costs, and if traffic growth is greater than predicted then operational problems may arise with queueing vehicles on the roundabout entries. For this reason the three-level roundabout is particularly suitable when right-turning flows and the proportion of heavy vehicles in the turning flows are relatively low.

Interchanges

In contrast to grade-separated junctions, interchanges provide uninterrupted movement for vehicles travelling from one mainline carriageway to another mainline carriageway, via link roads using a succession of diverging and merging manoeuvres.

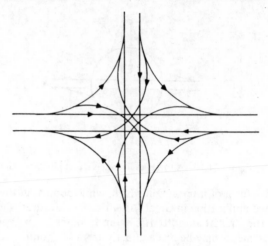

Figure 24.11   Direct connection between two primary distributors or motorways

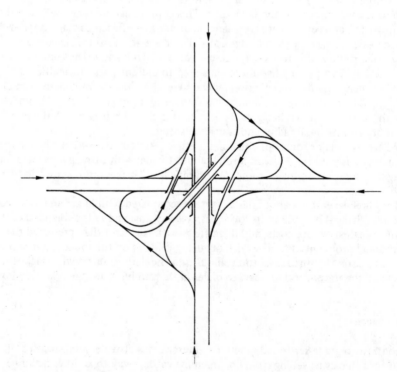

Figure 24.12   Three level interchange incorporating two loops

A form of interchange which has been used when high turning flows are predicted is the four-level layout shown in figure 24.11. This layout has a low landtake, relatively little structural content and does not contain loops. It is a form of interchange however which is considered visually intrusive and in addition has a large number of conflict points. An alternative layout to the four-level interchange is the three-level interchange incorporating two loops shown in figure 24.12. Whilst this interchange has fewer conflict points it is more expensive structurally than the four-level and is still considered visually intrusive. What is referred to as the cyclic or two level interchange is shown in figure 24.13; it uses reverse curve links to achieve a low number of conflict points. Its structural content is high and it requires extensive landtake but it produces little visual intrusion and is favoured where land is available.

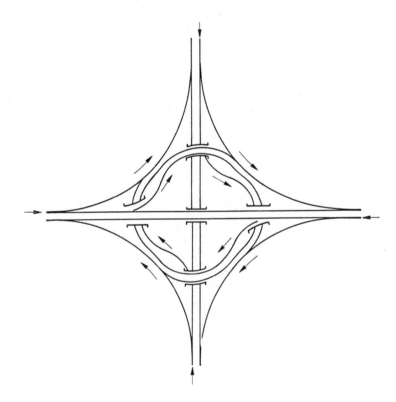

Figure 24.13   Two level cyclic interchange

Reference

1. Department of Transport, Layout of grade separated junctions, *Departmental Advice Note* TA 48/86, London (1986)

Problems

From the following junction types (a)–(i) select a layout that is likely to be satisfactory for the site and traffic conditions described in 1–6.
(a) A grade-separated rotary intersection with all the minor road flow passing
around the rotary section as shown in figure 24.14(a).

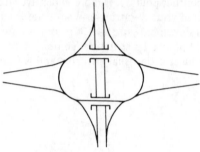

Figure 24.14(a)

(b) A grade-separated rotary intersection with only turning movements passing
around the rotary section as shown in figure 24.14(b).

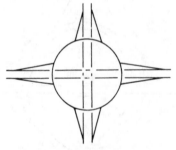

Figure 24.14(b)

(c) A single-level four-way signal-controlled intersection as shown in figure 24.14(c).

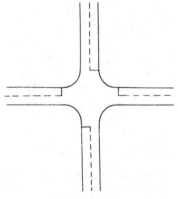

Figure 24.14(c)

(d) A single-level four-way roundabout intersection as shown in figure 24.14(d).

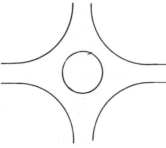

Figure 24.14(d)

(e) A semi-cloverleaf intersection as shown in figure 24.14(e).

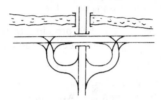

Figure 24.14(e)

(f) A diamond-type intersection as shown in figure 24.14(f).

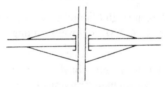

Figure 24.14(f)

(g) A trumpet intersection as shown in figure 24.14(g).

Figure 24.14(g)

(h) A semi-grade-separated roundabout as shown in figure 24.14(h).

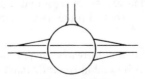

Figure 24.14(h)

(i) A grade-separated intersection with direct connections for all movements as shown in figure 24.14(i).

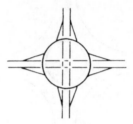

Figure 24.14(i)

1. Two major rural highways intersect at a right angled T-junction with heavy right turning movements from both highways. The traffic flows and accident potential are so great that grade separation is desirable.

2. In an urban area a heavily trafficked principal traffic route crosses a less heavily trafficked highway and there are a considerable number of turning movements between the highways.

3. Two motorways intersect in a rural area and it is desired to allow for a full range of turning movements between the two routes.

4. An all-purpose highway is crossed by a motorway but it is not anticipated that there will be many turning movements between the two highways. There are no topographical limitations on the design of the junction.

5. The junction is as described in (4) but a river runs parallel with the all-purpose highway preventing any entry or exit highways on one side of the all-purpose highway.

6. A right angled T-junction is to be constructed between a motorway and an all-purpose highway and it is anticipated that in the future the all-purpose highway will be extended so that a four-way connection will be required.

Solutions

The junction layouts that would be chosen for the given situation are:

1. At this intersection grade separation is desirable and as future extension of the three-way layout is not considered, then the trumpet intersection (g) would appear suitable.

2. As this intersection is heavily trafficked and there is a considerable number of turning movements then grade separation is desirable. It is an urban situation where land is likely to be limited and so a grade-separated roundabout (a) with all the traffic from the less heavily trafficked route passing around the roundabout would be used.

3. At the intersection of two motorways where provision for all turning movements has to be made then it is usual in Great Britain to use a grade-separated rotary

intersection (b) where a relatively compact layout is required or an intersection with direct connections (i) where more space is available or where speed reductions for the turning movements are undesirable.

4. Where an all-purpose highway crosses a motorway and there are few turning movements and no topographical limitations then a diamond intersection (f) is usual.

5. Where a diamond intersection cannot be provided because of physical obstructions then a semi-cloverleaf intersection (e) allows all the slop roads to be provided on one side of the minor road.

6. At a T-junction between a motorway and an all-purpose highway where it is required in the future to convert the T-junction into a four-way junction then a semi-grade separated roundabout (h) may be preferred. This layout allows the easy extension of the minor road into a four-way junction.

25

The environmental effects of highway traffic noise

Noise; measurement of sound levels

In our industrial society the number of sources of sound are steadily increasing and when these sounds become unwanted they may be classed as noises. Sound is propagated as a pressure wave and so an obvious measure of sound levels is the pressure fluctuation imposed above the ambient pressure.

If the graph of pressure against time for a single frequency is examined it is found to have a maximum amplitude (P_m) and in sound pressure measurements it is the root-mean-square pressure that is recorded. Some sounds are a combination of many frequencies while others are composed of a continuous distribution of frequencies. When this occurs the root-mean-square pressure values of all the individual frequencies are added together.

Using pressure units to describe sound levels requires a considerable range of numbers. It is frequently stated that the quietest sound that most people can hear has a sound pressure level of approximately 20 micropascals ($\mu N\ m^{-2}$) while at 100 m away from a Saturn rocket on take-off the sound pressure level is approximately 200 kPa. Rather than use a measurement system with this range, the ratio of a sound pressure to a reference pressure is used so that the sound pressure level is given in decibels by the ratio

$$20 \log_{10} \frac{\text{pressure measured}}{\text{reference pressure}} \quad \text{decibel (dB)}$$

The reference pressure is taken as 20 μPa.

Some idea of the range of sound pressure levels measured in decibels can be obtained from the values given in table 25.1.

When sound pressure levels are measured adjacent to a highway, a meter measuring in dB might indicate the same value when a fast moving motor cycle with a high pitched or high-frequency engine note passes and when a slow moving goods vehicle passes with a lower frequency note. The reason why the high pitched note is usually found more annoying than the lower one from the goods vehicle is that the human ear is more sensitive to sounds with higher frequencies than it is to sounds with lower frequencies.

222

TABLE 25.1   Some typical sound pressure levels
expressed in dB

Sound	Approximate sound pressure level (dB)
pneumatic drill	120
busy street	90
normal conversation	60
quiet office	50
library	40
quiet conversation	30
quiet church	20

If a sound level scale is going to be useful for measuring annoyance to human beings it must take this effect into account. Such a scale is measured in dB(A), the sound level measurements then being obtained by an instrument that weights the differing frequency components according to the curve given in figure 25.1. This

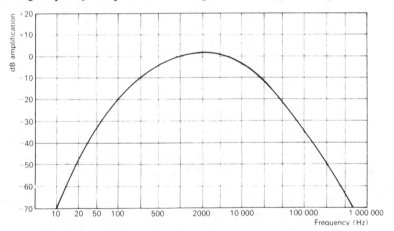

Figure 25.1   The 'A' weighting curve for sound-level meters

results in those frequencies that are relatively high or low receiving less weighting than those in the range 1 to 4 kHz. With a sound level meter reading in dB(A) it would therefore be found that the higher frequency note of the fast motor cycle would give a higher reading in dB(A) than the goods vehicle, although both produce the same sound pressure level measured in dB.

Sometimes it is necessary to know not only the sound level in dB or dB(A) but also the contribution that differing frequencies make to the overall sound. To obtain this information the sound is analysed by an instrument that passes it through a system of filters and allows the relative proportions to be determined.

Road traffic noise differs from most other sources of noise in that the level of noise varies both considerably and rapidly. If the variations of sound pressure level with time are recorded then a record of the type shown in figure 25.2 is obtained. At low sound-pressure levels the noise emitted from vehicles does not cause a great deal of annoyance but at higher levels the annoyance is considerable. For this reason many measures of noise nuisance specify a sound pressure level that is exceeded for 10, 20, 30 per cent etc. of the time.

When considering a scale that can be used to express a level of noise that should not be exceeded, it should be noted that this scale should be capable of expressing the relative effect on people of the noise being measured. Scholes and Sargent[1] have stated that the unit selected should meet at least the following requirements. Firstly the unit should correlate reasonably well with the criterion of dissatisfaction chosen so that noise levels measured using the unit will represent subjective reactions. Secondly a reasonably accurate set of design rules should be available, covering the estimation of noise exposure from traffic data and the estimation of the performance of noise control techniques, in terms of the chosen unit.

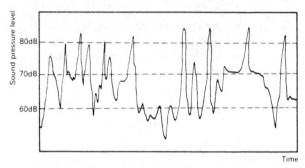

Figure 25.2 The variation of sound-pressure level with time for highway traffic

The London Noise Survey[2] measured noise levels at 540 sites and the subjective reactions of 1300 residents. During this survey particular values noted were the L_{10} level (the sound pressure level in dB(A) exceeded for 10 per cent of the time) and the L_{90} level (the sound pressure level in dB(A) exceeded for 90 per cent of the time). These levels represent the extremes of the range that were recorded for 80 per cent of the time. These values are referred to as the 'noise climate' and, together with the L_{10} value, are often quoted in the Wilson Committee Report[3].

Using data from interviews with 1200 residents at 14 sites in the London area where the roads were all straight, level and carrying free-flowing traffic, Langdon and Scholes[4] developed the 'Traffic Noise Index' (TNI). They found this index correlated well with dissatisfaction with noise conditions when the TNI was given by

$$4(L_{10} - L_{90}) + L_{90} - 30$$

A study carried out in Sweden[5] showed a good correlation between noise disturbance and three measures of noise, L_{10}, L_{50} and a noise exposure index based on the energy mean of the noise level, and given by the expression

$$L_{eq} = K \log \frac{1}{100} \Sigma 10^{L_i/K f_i}$$

where K is an empirically determined constant,
     L_i is the median sound level for the 5 dBA interval i
     f_i is the percentage time that a sound level is in the ith interval.

This unit has not however been found to correlate well with experience in the U.K. The difference in performance between the two countries is considered to be caused by the lower noise levels and the greater variability experienced in Swedish traffic conditions.

Another unit has been proposed[6] to cover a range of noise sources, whether highway traffic noise, aircraft noise or laboratory noise. It is referred to as the 'Noise Pollution Level' (LNP) and is given by the following expression

$$\text{LNP} = L_{eq} + 2.56\sigma$$

where L_{eq} is the energy mean noise level of a specified period
σ is the standard deviation of the instantaneous sound level considered as a statistical time series over the same specified period.

It has been found that the 'Noise Pollution Level' can express annoyance with traffic noise as well as the 'Traffic Noise Index' but both of these units suffer from the fact that their prediction under a wide range of circumstances is still uncertain. For this reason the Building Research Station[1] has proposed that, as an interim measure until further investigations are completed, the average L_{10} taken over the period 6 a.m. to 12 midnight on a weekday would provide a suitable standard for measuring traffic noise nuisance in dwellings. In a few years time however it is expected that sufficient experience will have been gained to use a unit incorporating the variability of traffic noise.

This unit is used to describe the noise exposure of a dwelling and is measured at 1 m from the mid-point of the facade of the building. It is the arithmetic average of the hourly levels in dB(A) just exceeded for 10 per cent of the time. These hourly values are obtained from sampling. The duration of each sample should include the passage of at least 50 vehicles and preferably 100 vehicles.

A maximum acceptable level of noise nuisance

The setting of an acceptable or a maximum level of noise presents considerable difficulties in compromising between what is desirable and what is physically and economically possible. The Wilson Committee[3] made a number of recommendations in 1963 in terms of L_{10} levels averaged over a 24-hour period. The maximum L_{10} levels inside buildings recommended by this Committee are given in table 25.2.

TABLE 25.2   Wilson Committee recommendation for maximum L_{10} levels indoors

Situation	Day	Night
country areas	40 dBA	30 dBA
suburban areas	45 dBA	35 dBA
busy urban areas	50 dBA	35 dBA

Scholes and Sargent[1] have suggested that as an interim standard a value of L_{10} (6 a.m.–midnight) of 70 dB(A) at residential facades should not be exceeded. The Noise Advisory Council has also recommended that as an act of conscious public policy existing residential development should in no circumstances be subjected to a noise level of more than 70 dB(A) on the L_{10} index.

The prediction of noise levels

The major factors which influence the generation of road traffic noise are:

(a) the traffic flow;
(b) the traffic speed;
(c) the proportion of heavy vehicles;
(d) the gradient of the road;
(e) the nature of the road surface.

In addition the following factors influence the noise level at a reception point distant from the highway:

(f) attenuation of the sound waves due to distance between source and receiver and also due to ground absorption;
(g) obstruction of the sound waves by buildings or noise barriers;
(h) obstruction of the sound waves due to a restricted angle of view of the source line from the reception point;
(i) reflection effects.

When predicting traffic noise levels by the procedure given in the Department of Transport Memorandum, Calculation of Road Traffic Noise[7], the factors (a)-(e) are used to predict the basic noise level (in terms of the hourly L_{10} or the 18-hour L_{10}) and the factors (f)-(i) are used to modify the basic noise level to obtain the prevailing noise level at a reception point.

Initially the basic noise level in terms of the 18-hour L_{10} or hourly L_{10} noise level is determined by the 18-hour or hourly traffic flow for a normalised source to receiver distance of 10 m when the mean traffic stream speed is 75 km/h, there are no heavy vehicles in the flow and the roadway is level. The traffic flow (Q) is the two way flow from 06.00 h to 24.00 h or the hourly flow (q) except when the two

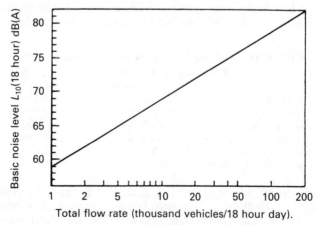

Figure 25.3   Basic noise level L_{10} (18 h)

carriageways are separated by more than 5 m or where the heights of the outer edges of the two carriageways differ by more than 1 m. In these cases the noise level of each carriageway is evaluated separately.

The relationship between L_{10} (18 h) basic noise level and traffic volume is shown graphically in figure 25.3 and expressed mathematically as

$$L_{10} \text{ (18 h)} = 29.1 + 10 \log Q \text{ dB(A)}$$

also L_{10} (hourly) $= 42.2 + 10 \log q$ dB(A)

A correction has to be made for the mean traffic stream speed and the percentage of heavy vehicles where these differ from 75 km/h and zero per cent respectively.

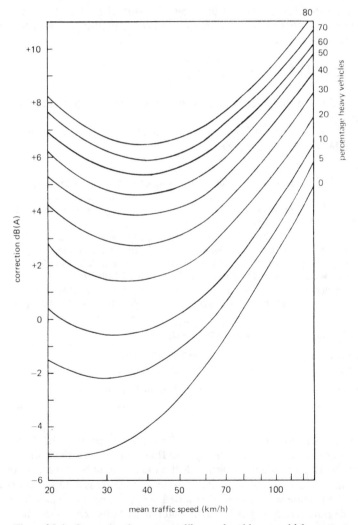

Figure 25.4 Correction for mean traffic speed and heavy vehicle content

The correction is shown in figure 25.4. Mathematically this correction is given by:

$$\text{Correction} = 33 \log \left(V + 40 + \frac{500}{V} \right) + 10 \log \left(1 + \frac{5P}{V} \right) - 68.8 \text{ dB(A)}$$

where P is the percentage of heavy vehicles in the flow (a heavy vehicle is any vehicle, other than a motor car, the unladen weight of which exceeds 1525 kg).

The traffic speed (V) may be obtained in either of two ways. It may be the prescribed highest mean speed in any one year within a 15-year period as given in table 25.3, or where local conditions indicate a significantly different value from the prescribed mean speed, then the highway authority may estimate the highest mean speed in a 15-year period.

When the speed is estimated from table 25.3 then the value of speed must be reduced because of the effects of heavy vehicles and gradient. This reduction in speed can be estimated from

$$\Delta V = \left[0.73 + \left(2.3 - \frac{1.15P}{100} \right) \frac{P}{100} \right] \times G \text{ km/h}$$

where P and G are the percentage of heavy vehicles and the percentage gradient respectively.

A correction equal to $0.3G$ has also to be made for the additional noise generated by vehicles on a gradient.

TABLE 25.3   Prescribed highest mean speeds in any one year within 15 years

Highway type		Prescribed speed
Special roads (rural) excluding slip roads	(speed limit less than 60 mph)	108 km/h
Special roads (urban) excluding slip roads	(speed limit less than 60 mph)	97 km/h
All-purpose dual carriageways excluding slip roads	(speed limit less than 60 mph)	97 km/h
Single carriageways, more than 9 m wide	(speed limit less than 60 mph)	88 km/h
Single carriageways, 9 m wide or less	(speed limit less than 60 mph)	81 km/h
(Slip roads are to be estimated individually)	(speed limit less than 60 mph)	
Dual carriageways	(speed limit 50 mph)	80 km/h
Single carriageways	(speed limit 50 mph)	70 km/h
Dual carriageways	(speed limit < 50 mph > 30 mph)	60 km/h
Single carriageways	(speed limit < 50 mph > 30 mph)	50 km/h
All carriageways	subject to a speed limit of 30 mph or less	50 km/h

carriageways are separated by more than 5 m or where there is one-way traffic then the correction only applies for the upward flow. If there is a single direction down gradient the Memorandum recommends the use of sound level measurements.

Road surface texture has a number of effects on noise generation depending on whether the texture is randomly distributed as is the case with many bituminous surfaces or transversely aligned as with concrete surfaces. The extent to which water can drain through a bituminous surface also has an effect on noise generation.

For roads which are impervious to surface water and where the traffic speed used in figure 25.4 is greater than 75 km/h the following corrections should be made. Correction for concrete surfaces is $10 \log(90TD + 30) - 20$ dB(A), for bituminous surfaces it is $10 \log(20TD + 60) - 20$ dB(A) where TD is the texture depth. Where road surfaces do not meet these requirements because the speed is less than 75 km/h then for impervious surfaces 1 dB(A) should be subtracted from the basic noise level. For roads surfaced with pervious macadams 3.5 dB(A) should be subtracted from the basic noise level.

With the basic noise level determined it is necessary to make corrections for the factors which affect the propagation of sound between the source and the reception position.

First a correction is made for the distance between the source and reception position. The nature of the ground influences distance attenuation and the Memorandum divides the ground over which the sound is propagated into hard ground and grassland. Hard ground is defined as mainly level ground, the surface of which is predominantly (more than 50 per cent) non-absorbent, that is, paved, concrete, bituminous surfaces and water.

The Memorandum gives the distance attenuation in these circumstances as:

$$\text{distance correction} = - 10 \log (d'/13.5) \text{ dB(A)}$$

where d' = the minimum slant distance between the effective source position and the reception point.

The effective source line is assumed to be 3.5 m in from the near kerb and at a height of 0.5 m above the road surface. Where the carriageways are considered separately then the source line for the far carriageway is taken as 3.5 m in from the far kerb and the distance from the kerb to be used in figures 25.5 and 25.6 is taken as 7 m in from the far kerb.

The distance correction is obtained graphically from figure 25.5 and in this chart the distance is measured from the edge of the nearside carriageway.

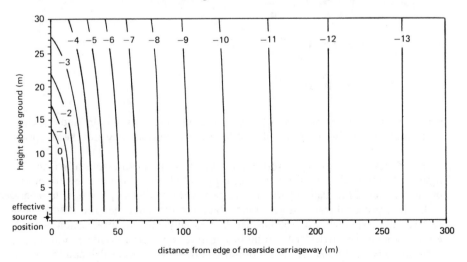

Figure 25.5 Correction for propagation over hard ground

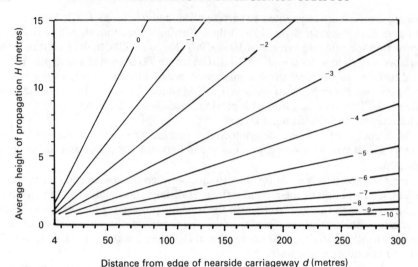

Distance from edge of nearside carriageway d (metres)

Figure 25.6   Correction for ground absorption, proportion of absorbent ground 1

Where the surface between the source line of the noise and the reception point is predominantly of an absorbent nature, such as grass, cultivated or planted land, the effects of ground absorption must be considered in addition to distance attenuation.

Mathematically the correction is given as:

$$\text{Correction} = 5.2I \log \left(\frac{6H - 1.5}{d + 35} \right) \ \text{dB(A)}$$

$$(d \geqslant 4 \ \text{m}) \quad 0.75 \leqslant H < \frac{d + 5}{6}$$

$$= 5.2I \log \left(\frac{3}{d + 3.5} \right) \ \text{dB(A)}$$

$$= 0 \ \text{when} \ H \geqslant \frac{d + 5}{6}$$

where d' is the slant distance between source line and reception point and d is the horizontal distance between the edge of the nearside carriageway and the reception point.

This distance and ground absorption correction is obtained graphically from figure 25.6 and in this chart the distance is measured from the nearside kerb or as previously stated for roads where the two carriageways are being considered separately.

In many practical cases the screening effect of objects between the source line and receiver must be taken into account. Dealing initially with the attenuation due to either a long noise barrier or a continuous obstruction caused by site conditions, the basis of the correction is the path difference between the source line and the reception point as illustrated in figure 25.7.

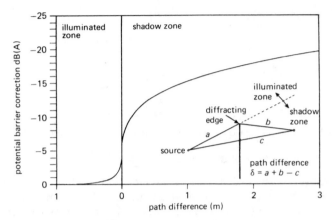

Figure 25.7   Potential barrier correction

This correction is applied to the basic noise level which has been corrected for distance using the hard ground correction. Ground absorption is ignored since the near ground rays are obstructed by the barrier.

Where only part of the road is shielded, for example by a short barrier, then a modified correction procedure has to be used which is conveniently illustrated by the following steps:

(a) Let θ_H be the total angle of view of the unscreened section of the source line for which the ground between the road and the reception point is hard. Let θ_S be the similar angle of view for which the ground is soft (grassland). Let θ_B be the total angle of view obtained by barriers. (Normally for straight roads $\theta_B + \theta_H + \theta_S = 180°$)

(b) Calculate the contribution to the noise at the reception point due to those lengths of line source covered by θ_H using the hard ground propagation correction. Correct this for the restricted angle of view by applying the correction $10 \log (\theta_H/180)$ dB(A).

(c) Repeat using θ_S and the grassland propagation and absorption correction and restricted field of view correction to obtain this contribution at the reception point. Combine with the value calculated in (b) to obtain L_U.

(d) Calculate the contribution from the screened section of road by calculating the unobstructed noise level at the reception point for hard ground and apply the long barrier correction and the restricted angle of view correction to give L_B.

(e) Combine L_U and L_B to give L_{10}.

In many situations reflection of noise from noise barriers or substantial buildings beyond the traffic stream along the opposite side of the road increases the noise level by $+ 1.5 (\theta' + \theta)$ dB(A) where θ' is the sum of the angles subtended by the reflecting surfaces.

These values calculated by these techniques are 'free field' noise levels. To calculate the noise level 1 m from a facade, as is required by the 1975 Noise Insulation Regulations, a correction of $+2.5$ dB(A) has to be made to allow for reflection from the facade.

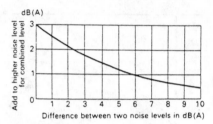

Figure 25.8   Combining exposures from two sources (adapted from ref. 1)

Controlling traffic noise by means of screens

When noise exposures from highway traffic are predicted it is frequently found that
the predicted exposures are greater than the recommended maximum values. A
method of reducing the noise exposure is by means of screens, making possible a
reduction in noise exposure averaging 10 dB(A).

A disadvantage of screening is the size of the barriers required because if noise
attenuation is to be obtained the facade being protected must be well within the
sound shadow formed by the screen. It should not be possible to see over the top or
around the ends of the screen if effective insulation is to be achieved. Narrow belts
of trees and shrubs are considered relatively unsatisfactory by the Building Research
Station for noise attenuation purposes. To be effective the belt should be about
50 m wide, dense and extend to ground level. The foliage should also be evergreen
for all-year screening.

To produce an effective sound shadow a noise barrier should either be close to
the highway or the building facade being protected. It should be dense enough to
create an effective shadow—a recommended value is at least 10 kg/m^2—and there
should not be any sound paths either through or under the barrier. Structural and
aesthetic considerations are also very important and these latter factors will obviously
influence the design.

Insulation of dwellings does however offer greater possibilities for the reduction
of noise nuisance than can usually be obtained in existing situations by distance or
noise barriers. It does not of course reduce noise nuisance in gardens and adjacent
areas.

The use of thicker glass has only a marginal insulating effect (approximately 1–3
dB(A) improvement) but double windows with a space of at least 150 mm between
the panes will give a sound insulation of between 20 and 30 dB(A) when one leaf is
sealed and the other is well fitting. Unfortunately the insulation value is reduced
when a powered fan is used to provide ventilation.

Controlling vehicle noise

The obvious way in which traffic noise may be reduced is by a reduction in the
noise emitted by individual vehicles and, as a consequence, the Quiet Heavy
Vehicle Project[8] was initiated under the overall direction of the Transport and
Road Research Laboratory. The project was a programme of research and development which aimed to produce a quiet diesel-engined heavy commercial vehicle
having external noise levels approximately similar to those of the private car.
Initially there was a research phase during which the various noise-producing com-

ponents of a standard vehicle were quietened experimentally, and later a development phase during which the 'research' vehicle was developed by industry into a commercially viable version for demonstration.

The standard vehicle selected for research was the diesel-engined heavy articulated vehicle as this was considered to be the most difficult to quieten, having the most powerful and noisiest engine; the upper limits of sound levels to be emitted by the vehicle components are given in table 25.4.

TABLE 25.4   Upper limits of sound levels to be emitted by vehicle components (from *Transport and Road Research Laboratory Supplementary Report* 746)

Source	Maximum level dB(A)	
	at 1 m	at 7.5 m
Engine and gear box	92	77
Air intake, exhaust system	84	69
Cooling system	84	69
Development of a practical exhaust system	84	69
Cab noise	75	

As a result of this programme, two vehicles were produced which emitted noise levels 10 dB(A) less than the original production vehicles and achieved similar reductions in internal cab noise. One was a standard Leyland Buffalo 4 × 2 tractor rated for operation at 32.5 tonnes and powered by a Leyland 510 158 kW (212 bhp) turbo-charged diesel engine. The other was a Foden/Rolls Royce tractor which has been developed by the vehicle and engine manufacturers into a fully engineered, practical and commercially viable vehicle which has met the target projects with a weight penalty of less than 1 per cent of the weight of a fully laden tractor-trailer combination.

The developed Foden quiet heavy vehicle was placed with a haulage contractor for a two year trial period. The object of the trial was to determine the durability of the vehicle's noise reduction features and to assess any effects which the reduction measures might have on the costs and difficulty of operating and servicing the vehicle. Data were collected on journeys made, loads carried and full consumption together with details of maintenance costs and problems of operation.

During the test period the vehicle travelled over 116 000 km and carried over 11 000 tonnes of payload, the maximum payload being carried on approximately 25 per cent of all journeys. The vehicle performed well with fuel consumption within the normally expected range; driver acceptability was good, partly because of the lower cab noise levels. It was reported that the vehicle received consistently good characteristics and the exceptionally quiet idling condition was particularly noted.

References

1. W. E. Scholes and J. W. Sargent, Designing against noise from road traffic, *Bldg Res. Stn Current Paper* CP.20/71
2. A. G. McKennel and E. A. Hunt, Noise annoyance in Central London. The Government Social Survey, *C.O.I. Report* SS.332 (1962)
3. Sir Alan Wilson, Chairman, *Noise–Final report of the committee on the problem of noise*, Cmnd 2056, HMSO, London (1963)
4. F. J. Langdon and W. E. Scholes, The traffic noise index: a method of controlling noise nuisance, *Bldg Res. Stn Current Paper* CP.38/68
5. Statens Institut For Byggnasforskning, Trafikbuller i Bosladsomraden, *Statens Institut for Byggnasforskning*, Rapport 36/68, Stockholm (1968)
6. D. W. Robinson, An outline guide to criteria for the limitation of urban noise, *Ministry of Technology, NPI, Aero Report Ac. 39* (1969)
7. Department of Transport, Welsh Office, *Calculation of Road Traffic Noise*, HMSO, London (1988)
8. J. W. Tyler and J. F. Collins, TRRL Quiet Heavy Vehicle Project, *Transp. Rd Res. Laboratory Report* LR 1067 (1983)

Problems

1. The sound pressure in the driving cab of a heavy goods vehicle is 2 Pa. Is the sound pressure level measured in dB, with reference to 20 μPa: 80, 100 or 150?

2. A heavy goods vehicle travelling at 60 km/h is found to produce a noise with a very low frequency in the range 100–500 Hz while a high-performance car travelling at 120 km/h is found to produce a noise with a frequency in the range 1000–2000 Hz. If both vehicles produce a noise with the same pressure level, which noise source will register the higher noise pressure level when measured in dB(A)?

3. Variations of the sound pressure level with time at a site adjacent to the highway are shown in figure 25.9. Indicate on the diagram which of the broken horizontal lines is likely to represent the L_{10}, L_{50} and L_{90} sound pressure levels.

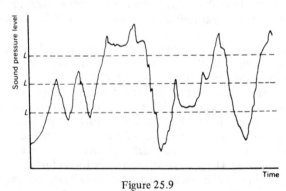

Figure 25.9

4. Many indices of traffic noise have been developed. From the given list of indices select the appropriate index to meet the requirements listed.

Indices
(a) The traffic noise index;
(b) The sound pressure level in dB(A) exceeded for ten per cent of the time;
(c) The noise pollution level;
(d) The sound pressure level in dB(A) exceeded for ninety per cent of the time.

Requirements
(a) An index that has been extensively correlated with the dissatisfaction of human beings with traffic noise;
(b) An index that can be used to express the noise nuisance of both aircraft and highway traffic.
(c) An index that has limited correlation with dissatisfaction with highway traffic noise but which expresses the variability of sound pressure levels;
(d) An index that expresses the background noise level.

5. It has been suggested that the L_{10} noise index should not exceed a given value at building facades. Is this value in the region of 30 dBA, 70 dBA or 100 dBA?

6. Calculate the noise exposure level at 1 m from the facade of a building 45 m from the nearest edge of a free-flowing traffic lane. The average commercial vehicle content of the traffic stream is 40 per cent; the mean stream speed is 60 km/h; the traffic volume is 15 000 vehicles/day. A second free-flowing traffic stream is 50 m distant from the building facade; there are no commercial vehicles in this stream; the mean stream speed is 80 km/h; the traffic volume is 30 000 vehicles/day. Ignore the effect of ground attenuation, gradient and road surface texture.

7. Calculate the improvement in noise exposure that will be obtained by the use of a very long barrier adjacent to a motorway where the geometrical layout is as shown in figure 25.10.

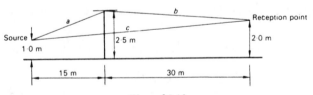

Figure 25.10

8. Calculate the improvement in noise exposure that will be obtained in the previous example if the barrier has a length of 200 m downstream and 50 m upstream of the reception point at which the noise exposure is to be calculated (figure 25.11).

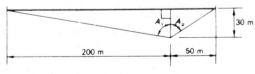

Figure 25.11

Solutions

1. Sound pressure level = $20 \log_{10} \dfrac{\text{pressure measured}}{\text{reference pressure}}$ dB

$$= 20 \log_{10} \frac{2}{2 \times 10^{-5}} \text{ dB}$$

$$= 20 \times 5 \text{ dB}$$

$$= 100 \text{ dB}$$

2. Reference to figure 25.1 shows that sounds with frequencies in the range 100–500 Hz produce less effect on the human ear than sounds in the frequency range 1000–2000 Hz. If both sounds have the same sound pressure level when measured in dB then the high-performance car with a noise in the frequency range 1000–2000 Hz will register a higher sound pressure level in dB(A).

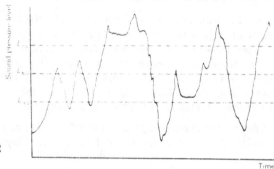

Figure 25.12

3. The correct placings of the L_{10}, L_{50} and L_{90} sound pressure levels are marked in figure 25.12.

4. The correct combination of indices and requirements is shown in table 25.5.

TABLE 25.5

Index	Requirement
(a) traffic noise index	(c) an index that has limited correlation with dissatisfaction with highway traffic noise but expresses the variability of sound pressure levels
(b) L_{10}	(a) an index that has been extensively correlated with the dissatisfaction of human beings with traffic noise
(c) noise pollution level	(b) an index that can be used to express the noise nuisance of both aircraft and highway traffic
(d) L_{90}	(d) an index that expresses the background noise level

5. The Noise Advisory Council has recently recommended that the L_{10} index should be used for measuring traffic noise disturbance. It also recommended that as a conscious act of public policy existing residential development should not be subjected to a noise exposure level greater than 70 dB(A) on the L_{10} index unless remedial or compensatory action was taken by the responsible authority (Reported in Hansard, 24th June 1971).

6. Calculation of noise exposure level from stream 1.

From figure 25.3   For a flow rate of 15 000 vehicles/18 h day L_{10} (18 h) = 70.9 dB(A).

From figure 25.4   Correct for speed and heavy vehicle content. Correction = 4.7 dB(A).

From figure 25.5   Correct for distance over hard ground (assume height of reception point is 2 m). Correction = −5.5 dB(A).

Reflection effect at facade = +2.5 dB(A)
L_{10} (18 h) at facade due to stream 1 = 72.6 dB(A)
Similarly for stream 2.

From figure 25.3 L_{10} (18 h) = 73.9 dB(A).
From figure 25.4   Correction = 0.5 dB(A)
From figure 25.5   Correction = −6.0 dB(A).

Reflection effect at facade = +2.5 dB(A).
L_{10} (18 h) at facade due to stream 2 = 70.9 dB(A).

From figure 25.8   Combined L_{10} (18 h) at facade = 75 dB(A).

7. The improvement in noise exposure that will be obtained by the use of a very long barrier adjacent to a motorway is calculated from the geometrical layout shown in figure 25.13.

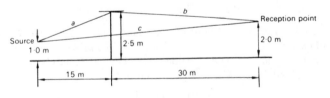

Figure 25.13

The improvement is related to $(a + b) - c$ and

$$a^2 = 15^2 + (2.5 - 1.0)^2$$

$$= 225 + 2.25$$

$$= 227.25$$

$$\therefore a = 15.074 \text{ m}$$

$$b^2 = 30^2 + (2.5 - 2.0)^2$$

$$= 900 + 0.25$$

$$\therefore b = 30.004 \text{ m}$$

$$c^2 = 45^2 + (2.0 - 1.0)^2$$

$$= 2025 + 1.0$$

$$= 2026$$

$$\therefore c = 45.011 \text{ m}$$

$$(a + b) - c = 0.067 \text{ m}$$

From figure 25.7 it can be seen that there is a reduction in the L_{10} noise exposure of approximately 9.5 dB(A).

8. The noise attenuation due to an unsymmetrical barrier may be calculated from a consideration of the relative geometric layout of the barrier and reception point.

$$\tan A_1 = \frac{200}{30} = 6.6667$$

$$A_1 = 81°28'$$

$$\tan A_2 = \frac{50}{30} = 1.6667$$

$$A_2 = 59°0'$$

$$\theta_B = 140°28'$$

Assume hard ground conditions, then $\theta_S = 0°$, $\theta_H = 39°32'$.
Let L_{10} be the noise level at the reception point when the barrier is not present. Correction for angle of view is

$$L_U = L_{10} + 10 \log (39°32'/180°) = L_{10} - 6.6 \text{ dB(A)}$$

Unobstructed noise level at the reception point corrected for barrier and field of view attenuation L_B is: $L_{10} - 9.5 + 10 \log (140°28'/180°) = L_{10} - 10.6 \text{ dB(A)}$.

Combining L_U and L_B gives shielded L_{10}. Then attenuation due to barrier is given by: shielded $L_{10} - L_{10} = (L_{10} - 10.6) + (L_{10} - 6.6)$

$$= 10 \log (10^{-1.06} + 10^{-0.66})$$

or, the attenuation of the barrier is 5.1 dB(A).

26

The environmental effects of highway traffic pollution

Air pollution from road traffic

During recent years there has been a widespread attempt to reduce air pollution from all sources. In the United Kingdom the Clean Air Act of 1956 has resulted in a noticeable decrease in coal consumption and a reduction in air pollution from domestic and industrial sources. During this same period there has been a marked increase in the volume of road traffic and consequently an increase in pollution from this source. The National Society for Clean Air[1] estimate that, during 1956 in the United Kingdom, the total coal and oil consumed was equivalent to 276 million tons (coal equivalent) and only 26 million tons of this were used for road or rail transport. It is however an increasing source of pollution, which is emitted in situations close to human activity. Approximately one-third of the carbon monoxide in the atmosphere is produced from vehicle exhausts.

The major sources of atmospheric pollution caused by motor vehicles have been given by Sherwood and Bowers[2] and may be classified as:

(a) exhaust gases;
(b) evaporative losses from the fuel tank and carburettor;
(c) crank case losses;
(d) dust produced by the wearing away of tyres, brake linings and clutch plates.

Considering the exhaust gases, the following compounds are normally present in the discharge from vehicle exhausts:

(a) carbon dioxide;
(b) water vapour;
(c) unburnt petrol;
(d) organic compounds produced from petrol;
(e) carbon monoxide;
(f) oxides of nitrogen;
(g) lead compounds;
(h) carbon particles in the form of smoke.

On occasions these components of the exhaust may react with each other to produce unpleasant secondary products. The most well-known effect of this type is the Los Angeles 'smog', which, because of the bright sunlight and the topography of the region, is formed by the reaction of the oxides of nitrogen and some of the hydrocarbons.

Both petrol and diesel engines give rise to similar products in their exhausts but the relative proportions differ. Diesel engine exhaust gases contain significantly lower proportions of pollutants than do those produced by petrol engines. Unfortunately, an incorrectly operated or maintained diesel engine is liable to emit smoke and produce an offensive smell but even then, apart from carbon particles, the degree of pollution is less than that produced by petrol engines.

The effects of these pollutants have been reviewed at the Transport and Road Research Laboratory[2] and the conclusions will be summarised.

Unburnt fuel and secondary products produced from the fuel

Unburnt fuel is emitted to the atmosphere by evaporation from the fuel tank and carburettor. A high proportion of the hydrocarbons in the crank case blow-by and in the exhaust gases also consists of unburnt fuel. The constituents of petrol are not considered to be toxic, but some of them have slight anaesthetic effects in high concentrations.

Many compounds are found in the gaseous products of the fuel emitted in the exhaust gases. Of these a significant proportion of aldehydes are produced. These have an irritant action on the eyes and on the respiratory system and they can be smelt in very small proportions.

In addition to the gaseous products a number of polynuclear aromatic compounds are also emitted with the exhaust gas in the form of very fine particles, which can persist in the air for lengthy periods. They are important because some of them, such as benzpyrene, are known to be carcinogenic, but the extent of the health hazard for the proportions present is not known. However, half the concentration of polynuclear aromatic hydrocarbons in the urban atmosphere is due to motor vehicles, the exposure level being equal to that produced by smoking one cigarette a day.

Carbon monoxide

The dangers of the absorption of carbon monoxide and its reaction with haemoglobin in the blood are well known. The degree of absorption depends on the carbon monoxide content in the air, the period of exposure and the activity of the individual. A survey by the Transport and Road Research Laboratory of the carbon monoxide content of the air in busy city streets in the United Kingdom has indicated that at the present levels, road users will not be aware of any discomfort from this source, but this may not be true for policemen and others operating in city streets for long periods of time.

While it is believed that carbon monoxide is unlikely to leave any permanent effects or cause acute physical discomfort, its effect cannot be entirely discounted because relatively small concentrations of carboxy-haemoglobin in the blood have been shown to temporarily impair mental ability. Fortunately this is only likely to occur in still weather in traffic jams, and even then only when the subject has been working hard for an hour.

It has been stated that an indication of the scale of the problem of carbon monoxide due to exhaust emissions is that cigarette smoking produces significantly higher exposure to carbon monoxide than that experienced by pedestrians on heavily trafficked roads. It should however be remembered that smoking in public places is likely to face increasing public opposition.

Oxides of nitrogen

Both nitric oxide and nitrogen dioxide are produced by the internal combustion engine, the former in much greater quantities than the latter, but nitric oxide oxidises to nitrogen dioxide. Typically in a city street there is twice as much nitric oxide as there is nitrogen dioxide.

Nitrogen dioxide is considerably more toxic than nitric oxide and a limited amount of data from around the most heavily trafficked roads in the United Kingdom show that the concentration exceeds the levels recommended in other countries. If an Air Quality Report is prepared then this pollutant should be investigated and likely concentrations exceeding 0.05 parts per million reported.

Lead compounds

In the United Kingdom a great deal of concern has arisen regarding the long-term hazard to health from lead due to vehicle exhaust emission. Airborne lead can be deposited on crops adjacent to highways and then enter the body via the food chain, although in most cases these crops make only a small contribution to the diet. Considering all sources of lead it has been recommended that the mean annual concentration of airborne lead should not exceed 2 micro-grammes per cubic metre in places where people might be continuously exposed for long periods. To achieve this level the lead content of petrol in the United Kingdom has been progressively reduced from (100) values of 0.4 grams per litre to 0.14 grams per litre, and it is expected that this will allow the recommended standard of 2 micro-grammes per cubic metre to be met in virtually all residential areas adjacent to major highways.

Estimation of exhaust pollution levels

It is generally accepted that the assessment of the air pollution due to a highway scheme may be made in terms of estimated levels of carbon monoxide[3]. Where it is desired to consider other pollutants this may be achieved by relationships between the levels of carbon monoxide and the concentrations of hydrocarbons, lead and oxides of nitrogen.

In the initial stages of road design it is usual for several alternative route proposals to be considered but only outline details of road centre lines and estimates of speeds and flows are available. A screening method for estimating air pollution has been developed to predict the annual maximum 8 hour concentration of carbon monoxide arising from traffic at a location near to a road network[4] which consists of straight roads, junctions and roundabouts. This method uses either a series of graphs or mathematical relationships obtained from a computer model which considers vehicle flow, vehicle speed and the distance of the point being considered from the roads. Meteorological and other variables have a large effect on the concentrations of pollutants but they are not considered in this screening process because the method provides an estimate of the maximum concentration likely to occur. The resultant chosen is the highest probable value from a distribution of 8 hour average concentrations which is based on the average hourly value.

The method makes use of four relationships. Firstly the connection between the concentration of carbon monoxide as a function of the distance D (m) between the road and the receptor, for long straight roads this is

$$C = 1.5 \exp(-0.025D)$$

This relationship was derived for A and B roads, a four-lane dual carriageway and a six-lane motorway. The fall of concentration with increasing distance from the road was found to be represented by this relationship when distance was measured from the centre line of each road type. The relationship was derived for a flow of 1000 vehicles per hour and for other flows the concentration should be multiplied by the expected peak hour flow in thousands of vehicles.

The relationship was derived for input weather and traffic conditions and gives the 1-hour average concentration expected for vehicles travelling at 100 km/h. A correction factor for different vehicle speeds can be made from the relationship

$$F = 38.9S^{-0.795}$$

where S is the vehicle speed (km/h)

F is the ratio of the emission rate at speed S to the emission rate at 100 km/h

A similar relationship was derived for roundabouts with a range of diameters, and a common relationship for all diameters was found when distances D (m) were measured from the centre line of the circulating carriageway. As roundabouts have a wide range of diameters it is suggested that where the central island diameter is 10 m or less then the roundabout should be considered as a straight stretch of road. The relationship is

$$C = 1.55 \exp(-0.033D)$$

There are no United Kingdom recommendations on exposure limits for ambient carbon monoxide and as a consequence United States Federal Air Quality Standards have been suggested as suitable for interpretation of the results of this predictive method. They specify carbon monoxide concentrations of 35 and 9 ppm which should not be exceeded more than once a year for exposure periods of one and eight hours respectively. The Transport and Road Research Laboratory have noted from air pollution survey data that the eight-hour standard is more difficult to meet than the one-hour standard and for this reason the eight-hour standard of 9 ppm was selected to indicate if a more detailed air quality survey was required. The previous relationships give a 1-hour average value and so it was necessary to estimate the 8-hour average value from the previously derived 1-hour average from

$$C_8 = 1.19 + 1.85C_1$$

where C_8 is the 8-hour concentration exceeded once a year

C_1 is the peak 1-hour concentration

As this method was derived for very long straight roads it is necessary to divide the road network into as few continuous roads as possible to prevent over-estimation of air pollution; only those sections nearest to the reception point should be considered. For example, a four-arm roundabout is divided into three sections, one for each of the two intersecting roads, and the third section is the circulating carriageway of the roundabout. A five-arm roundabout would have one additional section caused by the fifth leg and in all cases the distance would be the shortest distance (not necessarily perpendicular) to the section.

If this preliminary procedure indicates that concentrations of carbon monoxide give rise to concern, then the Department of Transport advise that detailed air quality investigations should be carried out using a suite of computer programs developed by the Transport and Road Research Laboratory[3].

This suite of programs has three major parts. The first and fundamental part predicts the hourly average concentration of carbon monoxide likely at a particular location for given weather and traffic conditions. This result may be used in the remaining two parts for estimating the range of concentrations of carbon monoxide likely for a variety of averaging periods or it may be used to give an indication of likely levels of oxides of nitrogen, hydrocarbons or lead at the site being considered.

References

1. National Society for Clean Air, *Air pollution from road vehicles—a report by the technical committee of the National Society for Clean Air*, London (1967)
2. P. T. Sherwood and P. H. Bowers, Air pollution from road traffic—a review of the present position, *Rd Res. Laboratory Report* 325 (1970)
3. A. J. Hickman and V. H. Waterfield, A user's guide to the computer program for predicting air pollution from road traffic, *Transport Road Research Laboratory Supplementary Report* 806, Crowthorne (1984)
4. V. H. Waterfield and A. J. Hickman, Estimating air pollution from road traffic: a graphical screening method, *Transport Road Research Laboratory Report* 752, Crowthorne (1982)

Problems

Are the following statements correct or incorrect?

1. The major source of pollution adjacent to heavily trafficked motorways is the diesel-engined goods vehicle.

2. Traffic 'smog' is likely to occur in regions where vehicle mileage is considerable and there is a low incidence of sunlight.

3. The benzpyrene content of the air in highway tunnels gives cause for alarm because it is greater than is found in industrial areas.

4. The carbon monoxide produced by vehicle exhausts in busy city streets is not likely to cause ill effects to vehicle drivers under free-flowing traffic conditions.

5. The concentration of oxides of nitrogen produced by vehicle exhausts as found adjacent to the highway is less than 1 per cent of the maximum allowable concentration for an 8-hour exposure in industrial conditions and has no long term effects.

6. The discharge from vehicle exhausts contains a stable aerosol of lead halides, which gives cause for concern because it produces a concentration of lead in the air of city streets which is considerably greater than the average level in rural areas.

Solutions

1. This statement is incorrect. Diesel-engine exhaust gases contain significantly lower proportions of pollutants than those produced by petrol engines.

2. This statement is incorrect. Traffic 'smog' is caused by the reaction of oxides of nitrogen and some of the hydrocarbons in the presence of bright sunlight.

3. This statement is incorrect. The benzpyrene content of the air in highway tunnels is less than that found in industrial areas.

4. This statement is correct. A survey carried out by the Transport and Road Research Laboratory has indicated that the carbon monoxide content of air in busy city streets in the United Kingdom is not likely to cause discomfort to road users when traffic is flowing freely.

5. This statement is incorrect. While the concentration of oxides of nitrogen produced by vehicle exhausts is as stated, the long-term effects are unknown.

6. This statement is correct.

27

Traffic congestion and restraint

The motor car is an invention which, within half a century, has revolutionised our way of life. It has made possible a dispersal of dwellings far exceeding that of the railway age, it has offered a wide choice of employment situations and has increased the scope of recreational activities to a remarkable extent.

At the same time the growth of vehicle ownership and use, together with population increases and the attraction of human activity into urban regions has resulted in considerable problems. The polluting effect of vehicle exhausts, the noise associated with road vehicles and above all the demand for physical space, all make it necessary to control the use of vehicles in urban areas.

The demand for road space, especially in existing central town areas, will always be greater than the supply because even if the necessary financial resources were available there would be conflicting demands for the available land. The fact that demand for road space is greater than the supply results in traffic congestion. While traffic management and urban highway construction have their place in minimising congestion it is now generally accepted that, without the dispersal of town centre activities, the only solution at the present time is a greater emphasis on public transport.

If this transfer from individual to public transport is accepted as one part of a solution to the problems of traffic in towns, then it will be necessary to find some means of traffic restraint. At the present time congestion itself acts as a restraint, causing trips which would take place at congested periods to be made at other times, or by alternative non-congested modes, or the trips may not be made at all. Congestion is however an inefficient form of restraint in that the priority of service is first come, first served, regardless of the value of the trip to either the tripmaker or the community. It is inefficient in the use of resources and is detrimental to the environment adjacent to the facility.

There are three general ways in which restraint could be applied. Firstly the entry of vehicles to certain areas at certain times could be prohibited by administrative means. On a limited scale this is already frequently employed in the form of pedestrian precincts but its application on a wide scale would involve the entry of specialist service or emergency vehicles and would involve decisions as to who should be allowed entry on the grounds of the value of their trip either to themselves or to the community.

Secondly restraint could be applied by the use of parking regulations. The Ministry of Transport report, 'Better Use of Town Roads'[1], considered that the most promising method of restraint, at least for the shorter term, would be to intensify control over the location, amount and use of parking space, both on and off the street. They especially considered it necessary to restrict long-term parking, which is characteristic of car commuting. They felt that in some places control over the use of publicly available parking space might not be adequate and might have to extend to privately available parking space, even though this would be costly and require new legislation.

The third form of restraint that could be applied is road pricing. This is a form of road user taxation whereby users of congested roads would be charged according to the distance travelled or the time spent on them, at varying rates governed by the degree of congestion. A system of road pricing does however exist; it was introduced by Lloyd George in 1909 and basic features of the system are still unchanged. It is not however an efficient form of road pricing because the annual licence is not related to road use and petrol tax does not discriminate to any significant extent between tripmaking in congested and non-congested conditions. It has been said[2] that, 'the issue is not whether we should have a system of road pricing, because we already have one, but whether we could devise a better one'.

Road pricing is an attempt to change the principle that highways should be treated as welfare services, that is financed out of taxes, to the principle that they should be treated as public utility services for which charges are made. It has been argued[3] that road pricing is democratic because it is the tripmaker who makes the decision as to whether or not the trip should be made at the given price rather than a government body making a decision that his trip was in the interests of the community.

If the price is placed at a level that reflects the costs a tripmaker imposes on others, it will produce traffic flows that reflect social benefits and costs as well as the tripmaker's private benefits and costs. If alternative means of transport are also priced in this way then the resultant traffic flows would give a better indication of the need for the construction of future transport facilities. This is because the true demand for transport facilities will be known rather than relying on projections of 'free' tripmaking in which the tripmaker does not bear any of the cost imposed on others.

Some doubts on road pricing have been stated by Lichfield[4]. These are that conditions approximating to perfect competition are difficult to visualise for transportation systems. It is also difficult to estimate a price charge that reflects the indirect repercussions caused to external economies or diseconomies in production or consumption. When environmental costs have to be included as well as congestion costs then determination of the pricing charge requires further research. Finally income distribution is likely to be affected when a uniform pricing charge is imposed upon all tripmakers.

The Ministry of Transport report, 'Road Pricing: The Economic and Technical Possibilities'[5], gives the following requirements for a pricing system.

1. Charges should be closely related to the amount of use made of the roads.

2. It should be possible to vary the price for different roads, at different times of day/week/year or for different vehicle classes.
3. Prices should be stable and ascertainable by road users before they commence the journey.
4. Payment in advance should be possible although credit facilities should in certain cases be permissible.
5. It should be accepted as fair.
6. It should be simple.
7. Equipment used should be reliable.
8. The system should be reasonably free from fraud and evasion, both deliberate and unintentional.
9. If necessary it should be capable of being applied to the whole country.

They also considered that the following would be desirable, but not so important, requirements.

1. Payment should be in small amounts, at fairly frequent intervals, say £5/month (1964 prices).
2. Drivers in high-cost areas should be aware of the amount they are paying.
3. Drivers' attention should not be diverted by the pricing system.
4. The system should be applicable to drivers from abroad.
5. Enforcement should lie with traffic wardens.
6. If possible it would be preferable to use the method for charging for street parking.
7. It should give a measure of the strength of demand for road space, so as to guide the planning of new routes.
8. The system should be capable of gradual introduction starting with an experimental stage.

In addition to parking charges the Ministry Group examined the possibility of entry charges using a system of supplementary licences, its main advantage over parking charges being that it restrains both non-parking tripmakers and those who are able to park privately. Its disadvantage is that it fails to take into account different amounts of road space within the licensing area, or use at different levels of congestion. The first disadvantage limits the size of the licensing area, and in addition the larger the area, the higher will be the licence fee and the greater the effects at the boundaries. In 'Better Use of Town Roads' it is suggested that the maximum practicable area for such a system might be no more than 10 square miles.

Another indirect method of charging for road use which was examined was the employment of differential fuel taxes. A different petrol tax could be levied in different areas according to the congestion in the area. This tax however could be avoided by filling up in low-tax areas or it could be avoided by special fuel-carrying journeys unless the differential was small. This is only a general tax on a large area, including congested and uncongested roads in the same area at congested and uncongested times. The Report considered that it would affect individual garages, cause a black market in petrol and encourage people to carry cans of petrol. In addition commuters would find it easy to avoid tax unless the tax area was very large.

The Panel considered that electronic metering offered the best opportunity for meeting the requirements for a pricing system. They outlined seven systems, five of which were automatic and two were driver operated. A distinction was made between continuous systems in which the charge is made on the basis of time spent or distance travelled in the charging area, and point systems in which a driver is charged every time he passes a pricing point.

In subsequent development work the Transport and Road Research Laboratory[6] rejected driver-operated and continuous automatic systems. Their reasons for this decision were given as:

(a) Driver-operated systems impose an additional demand on the driver and may distract his attention. In addition they are vulnerable to attempts at evasion.

(b) Systems based on measuring time or distance require time references or distance-measuring methods to be used in the vehicle.

(c) There is the difficulty that a charge should not be registered when the vehicle enters private premises: this is a particularly difficult problem when a time-based system is used.

(d) The time-based system would result in an attempt to minimise time with a consequent decrease in safety.

(e) Continuous systems are not fail-safe in that failure to record the departure of a vehicle from the charging zone would result in excessive charges being levied.

These disadvantages may be avoided by the use of a point-pricing system, whose advantages are:

(a) Point-pricing systems are flexible because the charging rate can be varied by alteration in the spacing of the pricing points. The system can also be easily extended by the addition of further pricing points.

(b) Variations in the charging rate can be obtained by varying the price per charging point or by switching out series of points at non-congested times of the day.

(c) Relatively simple apparatus is required on the vehicle for this type of system.

(d) The point-pricing system is considered to be consistent with existing road charging facilities and could be an extension of a system of supplementary licences. Regular entrants to the pricing zone would begin to mount special equipment in their vehicles while infrequent entrants would purchase occasional daily licences. In some circumstances it would be possible to use the point-pricing system to levy parking charges.

(e) A point-pricing system is fail-safe in that if a vehicle fails to register passing a given point, the tripmaker gains.

The point-pricing system makes it possible to place pricing points anywhere on the road network but they should be placed so as not to result in undesirable route selection, preferably in the form of closed cordons. This would result in the charging area being divided into small zones with a pricing point imposing a charge every time a vehicle passes from one zone to another.

At the Transport and Road Research Laboratory two point-pricing systems are being developed. One is an on-vehicle system in which the meter that stores the number of pricing points passed is mounted on the vehicle. The other is an off-vehicle system in which each vehicle that passes a pricing point is uniquely detected and the information is then passed on to a central computer.

While the implementation of a full road-pricing scheme is obviously still far distant the report 'Better Use of Town Roads' considered direct pricing as potentially the most efficient means of restraint.

If any road-pricing system is to be introduced it will require considerable capital investment and incur substantial maintenance costs. It must also be shown that the net economic benefits of the scheme, that is the economic benefits minus the operating costs, are worth while when compared with other forms of restraint.

The variations of net benefit that occur with increasing traffic flow on a highway can be illustrated by the following example. Suppose that all tripmakers consider the value of a trip on this highway is 50 p. The tripmakers' private costs of the journey are normally composed of two elements. Firstly there are vehicle operating costs including fuel (but excluding tax), maintenance and depreciation which may be expressed as a cost per vehicle kilometre. The second element consists of those costs that vary with journey speed, and mainly depend on the cost of individuals' travel time. For this example the travel costs given in 'The Economic Appraisal of Inter-Urban Road Improvement Schemes'[7] will be used; they are reproduced as table 27.1. These travel costs comprise both the fixed and variable private costs.

It will also be assumed that the relationship between travel speed and flow for this highway has the form

$$v = 87 - \frac{q + 1400}{82} \qquad \text{or} \qquad 75 \text{ km/h} \qquad\qquad (27.1)$$

whichever is the least, where q is the flow in vehicles/hour.

The variation of costs and benefits as the flow along this highway link increases are calculated and tabulated in table 27.2.

It can be seen from table 27.2 and figure 27.1 that the net benefits from operating the system increase as the traffic flow increases until an optimum flow value, from the point of view of net benefits obtained, of approximately 3700 veh/h is reached.

The additional cost of adding one extra vehicle to the traffic stream is referred to as the marginal cost and in this example it is calculated by noting the change in

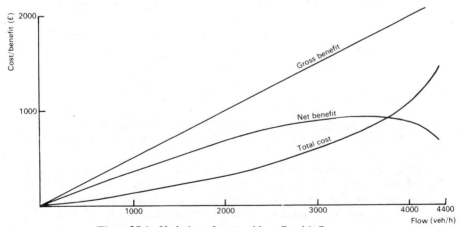

Figure 27.1   Variation of cost and benefit with flow

TABLE 27.1 Private travel costs at varying speeds

Vehicle speed (km/h)	Travel costs (p/km)		
17	6.78	54	2.88
18	6.45	55	2.85
19	6.16	56	2.83
20	5.91	57	2.80
21	5.68	58	2.78
22	5.47	59	2.75
23	5.28	60	2.73
24	5.11	61	2.70
25	4.96	62	2.68
26	4.82	63	2.65
27	4.69	64	2.63
28	4.56	65	2.60
29	4.44	66	2.58
30	4.32	67	2.56
31	4.21	68	2.54
32	4.11	69	2.52
33	4.01	70	2.51
34	3.93	71	2.49
35	3.85	72	2.47
36	3.78	73	2.46
37	3.70	74	2.44
38	3.63	75	2.42
39	3.56	76	2.41
40	3.50	77	2.39
41	3.44	78	2.37
42	3.38	79	2.36
43	3.32	80	2.34
44	3.26	81	2.33
45	3.20	82	2.31
46	3.15	83	2.30
47	3.10	84	2.29
48	3.06	85	2.28
49	3.02	86	2.27
50	2.99	87	2.26
51	2.96	88	2.25
52	2.91	89	2.25
53	2.90	90	2.24

costs and benefits when the flow increases by 100 veh/h. It should be noted that since the rate of change of costs is not uniform this will only give an approximate value.

From table 27.2 it can be seen that at low levels of flow the marginal or additional cost of one vehicle is considerably less than the benefit of the journey. At flows exceeding 3700 veh/h the marginal cost is greater than the benefit of the journey, and additional vehicles impose greater costs than the benefits they receive. These journeys would still be made however, because even at a flow of 4300 veh/h

TABLE 27.2 Variations of gross and net benefits with increasing flow

1	2	3	4	5	6	7	8
						Increase due to the addition of one vehicle	
Flow q (veh/h)	Speed equation (27.1) (km/h)	Cost/km vehicle (p)	Total cost (1) × (3) × 5 km (£)	Gross benefit 50 p × (1) (£)	Net benefit (5) − (4) (£)	Cost (£)	Net benefit (£)
100	69	2.52	12.60	50	37.40		
200	67	2.56	25.60	100	74.40	0.13	0.37
500	64	2.63	67.75	250	182.25		
600	63	2.65	79.50	300	220.50	0.14	0.36
1000	58	2.78	139.00	500	361.00		
1500	52	2.93	219.75	750	530.25		
1600	50	2.99	239.20	800	560.80	0.19	0.31
2000	46	3.15	315.00	1000	685.00		
2500	39	3.56	445.00	1250	805.00		
2600	38	3.63	471.90	1300	828.10	0.27	0.23
3000	33	4.01	601.50	1500	898.50		
3100	32	4.11	637.05	1550	912.95	0.36	0.14
3500	27	4.69	820.75	1750	929.25		
3600	26	4.82	867.60	1800	932.40	0.47	0.03
3700	25	4.96	917.60	1850	932.40	0.50	0.00
4000	21	5.68	1136.00	2000	864.00		
4100	20	5.91	1211.55	2050	838.45	0.76	−0.26
4300	17	6.78	1457.70	2150	692.30		

the cost/km to each driver is 6.78 p and the total journey cost is 34.90 p, which is less than the benefit of 50 p he would receive by making the trip.

In a real situation however not all tripmakers would attach the same value to the benefit they would obtain from the trip. Because of differences in income and in the nature of the trip as well as differences in the value of getting to the destination the demand for tripmaking may be expected to fall as the cost of the trip increases.

The demand D for tripmaking may be expressed as a function of the private cost or price p of the trip, that is

$$D = f(p) = Kp^{-\gamma}$$

where K and γ are parameters. The general form of this relationship is given in figure 27.2.

When the effects of varying trip costs are being considered an important factor in determining the relationship between a change in trip cost and the resulting change in tripmaking is the price or cost elasticity e_p. It is defined as the percentage change in quantity demanded that results from a 1 per cent change in price. Then

$$e_p = \frac{\partial D/D}{\partial p/p}$$

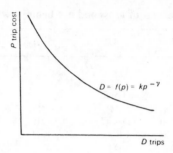

Figure 27.2   General form of the demand function

$$= \frac{p}{D} \times \gamma K p^{-\gamma-1}$$

$$= \frac{K\gamma p^{-\gamma}}{D}$$

$$= -\gamma$$

If the demand function is K/p, as is frequently assumed, then the price elasticity is unity.

The cost to the tripmaker has been referred to as the private cost of a journey while the cost that a tripmaker imposes upon other tripmakers because of an increase in congestion is referred to as the congestion cost.

As has been indicated previously the addition to the total costs caused by one extra tripmaker is the marginal cost. It consists of the private costs of the additional trip together with the congestion costs caused by the additional trip.

In addition to these costs there are environmental costs that the trip imposes upon the area adjacent to the highway, and road maintenance costs.

Private costs are divisible into two parts, those proportional to distance, such as fuel, maintenance, depreciation, and those varying with journey speed, which are chiefly associated with the value of time.

Congestion costs arise from the interaction of vehicles and they are illustrated in the typical relationship between speed and flow shown in figure 27.3, where q_1 is the flow at which interaction commences. It may be assumed that there is a linear

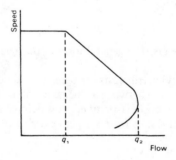

Figure 27.3   A general form of the speed/flow curve

relationship between q_1 and q_2, the value of maximum flow. This relationship may be assumed to be of the form

$$v = a - bq \qquad (27.2)$$

If it is assumed that the total private cost of a trip may be expressed as

$$c + \frac{d}{v} \text{ pence/km} \qquad (27.3)$$

where c is the component that is proportional to distance,
d is the component that is proportional to time,

then the private costs of a tripmaker are

$$c + \frac{d}{a - bq} \text{ pence/km}$$

so that the private cost/flow curve has the form shown in figure 27.4. The intersection of the demand curve and the private cost/flow curve at E represents the equilibrium condition at cost c_e and flow q_e. Additional trips above flow q_e will not be made because the private cost of the trip is greater than the benefit of the trip as given by the demand curve.

At this equilibrium condition the net benefit to tripmakers, that is the benefits minus the private costs, is represented by the vertically hatched area.

If however the highway is improved then it will result in a revised private cost/flow curve with a new equilibrium point B and the flow will be q_f at a trip cost c_f. In this case the net benefit to tripmakers will be given by the diagonally hatched area.

The marginal cost/flow curve may be obtained by differentiating the expression for the total cost of tripmaking.

$$\left(c + \frac{d}{a - bq} \right) q$$

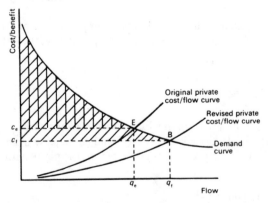

Figure 27.4 The change in net benefit to tripmakers due to a highway improvement

with respect to q, giving the equation of the marginal cost/flow curve

$$c + \frac{d}{v} \times \frac{a}{v} \qquad (27.4)$$

It can be seen that this is of the same form as the cost/flow curve with the private cost component multiplied by the factor a/v. From a knowledge of the speed/flow relationship a/v will be greater than unity and the marginal cost/flow curve will have the general relationship to the private cost/flow curve shown in figure 27.5.

In figure 27.5 the equilibrium flow and private cost is given by the intersection of the marginal cost/flow and demand curves at E, a situation which would occur when a decision as to whether a trip should be made is based solely on private cost. The intersection of the marginal cost/flow curve with the demand curve at O however gives the optimum flow. If flow is restricted to below this level then only trips with a marginal cost less than the value of the trip to the tripmaker will be allowed. If the flow is greater than the optimum then a trip is allowed that has marginal costs greater than the value of the trip.

If traffic flow conditions are shifted from the equilibrium position F to the optimum position O then the change in net benefit will be represented by the area $A-c_p-P-O$ minus the area $A-c_e-E$. This is in fact equal to the area c_e-c_p-P-T minus the area OTE, which represents the gain in net benefit enjoyed by the tripmakers who are not removed due to improved speed, minus the net benefit previously enjoyed by the tripmakers who are now removed.

The movement from E to O can be achieved by the use of a road pricing charge, represented by PO. Individual tripmakers will however be worse off under the pricing system because the decrease in private costs TP will always be less than the pricing charge OP.

An estimate of the economic benefits to be derived from direct road pricing is contained in 'Road Pricing: The Economic and Technical Possibilities'[5] and the following illustration of a method which may be used to calculate the net benefits from the imposition of a road-pricing charge is based on that report.

A highway network is considered where the flow is q and the corresponding private costs of a tripmaker are p. The demand curve is of the form $q = f(p)$, which may alternatively be expressed as the inverse relationship $p = f^{-1}(q)$, where $f^{-1}f(p) = p$.

Consider the situation before any price is imposed and let the private cost be α and the flow be Q. From the reasoning given previously, the net benefit to all tripmakers is

$$\int_0^Q f^{-1}(q)\, dq - \alpha Q \qquad (27.5)$$

Consider now the situation in which a road pricing charge per vehicle/km of β is imposed, causing a decrease in flow to Q^1 and an increase in speed and a reduction in costs from α to α^1 (let $\alpha - \alpha^1 = g$). As before the net benefit to tripmakers is now

$$\int_0^{Q^1} f^{-1}(q)\, dq - (\alpha^1 + \beta) Q^1$$

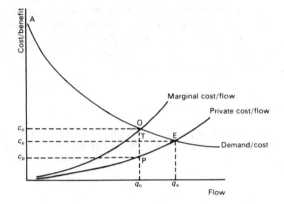

Figure 27.5 The relationships between marginal and private costs and flow, and between demand and cost

The amount βQ^1 is however a transfer payment and represents no real cost to the community, giving a net benefit in the road-pricing situation of

$$\int_0^{Q^1} f^{-1}(q) \, dq - \alpha^1 Q^1 \tag{27.6}$$

The increase G in net benefits from the imposition of the road charge β is then

$$G = \alpha Q - \alpha^1 Q^1 - \int_{Q^1}^{Q} f^{-1}(q) \, dq \tag{27.7}$$

If it is assumed that when the flow is of the order of Q or Q^1 the demand function is

$$q = f(p) = \frac{k}{p}$$

then the elasticity of demand is unity and

$$Q = \frac{k}{\alpha}, \quad Q^1 = \frac{k}{\alpha^1 + \beta}, \quad p = f^{-1}(q) = \frac{k}{q}$$

By substituting in equation 27.7 it can be shown that the increase in net benefits is

$$G = \beta Q^1 + \alpha Q \log \frac{Q^1}{Q} \tag{27.8}$$

and therefore

$$\frac{G}{Q} = \beta \frac{Q^1}{Q} + \alpha \log \frac{Q^1}{Q} \tag{27.9}$$

where Q^1/Q is the ratio of the new flow to the old flow.

In order to quantify the above relationship a process of inspection is necessary. because Q^1/Q varies with β. In the first place the relationship between g and Q^1/Q must be established from speed/cost and speed/flow relationships. The report took a speed/cost relationship derived by Charlesworth and Paisley[8] and updated it to values at the date of the report (1964) to give

$$\alpha = \frac{1}{13.92}\left(4.1 + \frac{224}{v}\right) \qquad (27.10)$$

where α was the average private cost in shillings for one passenger car unit travelling at v miles/h. The speed/flow relationship employed was one derived by Thomson[9] for Central London

$$v = 28 - \frac{q}{125} \qquad (27.11)$$

where q is the flow in p.c.u.s per hour and v is the speed in miles/h.

It was then possible to calculate the average private cost (from equation 27.10) and the corresponding flow (from equation 27.11) for a range of speeds. For the difference between any two journey speeds of which the lower is the congested or base speed the difference in the private costs (g) and the ratio of the two flows (Q^1/Q)may be tabulated.

For varying base or congested speeds it is then possible to insert values of Q^1/Q, corresponding values of α and varying values of β into equation 27.9 to obtain G/Q, the net benefits per vehicle mile. The variation of net benefits for differing pricing charges β derived[5] on this basis is illustrated in figure 27.6.

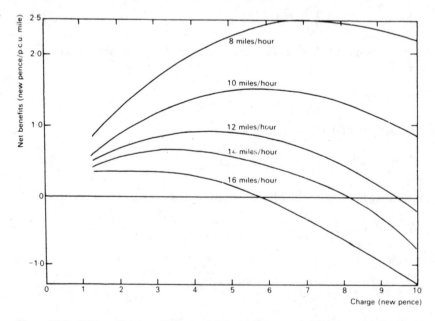

Figure 27.6 Net benefits from different charges at different speeds (adapted from ref. 5)

References

1. Ministry of Transport, *Better use of town roads*, HMSO, London (1967)
2. J. M. Thomson, Case for road pricing, *Traff. Engng Control*, (March 1968), 536-9
3. G. J. Roth, *Paying for Roads: The Economics of Traffic Congestion*, Penguin, Harmondsworth (1967)
4. N. Lichfield, Planner/economists view of road pricing. *Traff. Engng Control*, (February 1968), 485-7
5. Ministry of Transport, Road Pricing: The Economic and Technical Possibilities, HMSO, London (1967)
6. G. Maycock, Implementation of traffic restraint, *Report* LR 422, Transport and Road Research Laboratory (1972)
7. Department of the Environment, *The Economic Appraisal of Inter-Urban Road Improvement Schemes* (1971)
8. G. Charlesworth and J. I. Parsley, The economic assessment of returns from road works, *Proc. Instn Civ. Engrs*, **14** (1959), 229-54
9. J. M. Thomson, Calculations of economic advantages arising from a system of road pricing, *Paper PRP 18*, Transport and Road Research Laboratory (1962)

Problems

Select the correct completions to the following statements.

1. In the central areas of existing cities the future use of the private motor vehicle:

(a) will be made possible by higher parking charges;
(b) will be prohibited;
(c) will be assisted by the construction of major highway schemes and parking garages.

2. The report 'Better use of town roads' concluded that in the near future the use of private cars in the central areas of cities would be restrained by:

(a) an electronic road pricing system;
(b) the control of all parking facilities and the levying of considerably higher parking charges;
(c) regulating entry to certain areas by admission charges;
(d) the use of differential fuel taxes.

3. The object of road pricing is:

(a) to ensure that a journey would not be made if the cost the journey imposed on others was less than the benefits obtained by the tripmaker;
(b) to allow individual tripmakers to make their own decision as to whether to make a trip by private transport after taking into account the costs and benefits of the trip;
(c) to completely remove traffic congestion from the highway network.

4. A speed/flow relationship for a highway has the form shown in figure 27.7.

When the flow on the highway is:

(i) in the region $0-q_1$
(ii) in the region q_1-q_2

will the additional cost of introducing an extra vehicle into the flow be:

(a) the private costs of the additional vehicle?
(b) the private costs of the additional vehicle together with an addition to the private costs of other vehicles?
(c) negligible?

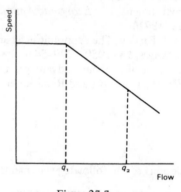

Figure 27.7

5. Indicate on figure 27.8 which curve is:

(a) a demand curve on which tripmakers all value the benefits of their trip equally;
(b) a demand curve on which some tripmakers value the benefits of their trip at a higher level than others;
(c) a cost/flow curve on which costs are a function of journey distance and time;
(d) the marginal cost/flow curve for the above cost/flow curve.

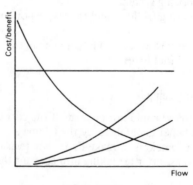

Figure 27.8

6. A highway link has the following characteristics:

(a) a speed/flow relationship of $v = 50 - q/100$
where v is the stream journey speed (km/h)
q is the stream flow (p.c.u./h)
(b) a flow/cost relationship of $p - 1.25(5 + 300/v)$
where p is the private cost in p.c.u./km
v is the journey speed in km/h.
(c) a demand relationship of the form

$$D = 4 \times 10^4 p^{-1}$$

where D is the demand in p.c.u./h
p is the private cost in km/h.

Determine graphically:

(a) the flow and the speed on the highway when each tripmaker is only aware of his private costs;
(b) the flow and the speed on the highway when a trip is only made if the benefit of a trip to the tripmaker exceeds the additional cost imposed on other tripmakers;
(c) the pricing charge that would result in the flow (b).

Solutions

1. (a) This statement is correct. The report 'Better Use of Town Roads' concluded that for the short term the most promising form of restraint is likely to be the intensification of controls over parking. It is this restraint over the less valued trips that will make it possible to use private motor vehicles for the more valued trips.

(b) This statement is incorrect. It would not be practicable to prohibit the entry of all private motor vehicles into larger town areas and entry by permit would raise administrative difficulties.

(c) This statement is incorrect. Large-scale highway and parking garage construction would be necessary on so large a scale that it is doubtful if the financial resources could be made available. If it were financially possible then it would result in the wholesale reconstruction of central city areas.

2. (a) This statement is incorrect. The report concluded that this method is potentially the most efficient form of restraint, but that it would take several years to develop.

(b) This statement is correct. The report concluded that the strict control of parking would in the short term be the most promising form of restraint.

(c) This statement is incorrect. The report concluded that they were far from satisfied that it would be practicable to regulate entry to certain areas by admission charges.

(d) This statement is incorrect. The report did not recommend the use of differential fuel taxes.

3. (a) This statement is incorrect. The correct statement is . . . to ensure that a journey would not be made if the cost of the journey imposed on others was greater than the benefits obtained by the tripmaker.

(b) This statement is correct.

(c) This statement is incorrect. Road pricing alone will not produce a cure for congestion because the need for investment in urban highways will still remain.

4. When the flow on the highway is (i) in the region $0-q_1$, the additional cost of introducing an extra vehicle will be the private costs of the additional vehicle, as stated in (a) because increasing flow does not result in a decrease in speed.

When the flow on the highway is (ii) in the region q_1-q_2, the additional cost of introducing an extra vehicle will be the private costs of the additional vehicle together with an addition to the private costs of other vehicles, as stated in (b), because increasing flow results in a decrease in speed.

5. The appropriate curves are as indicated in figure 27.9.

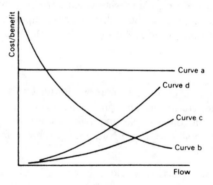

Figure 27.9

6. From equation (27.2), $v = a - bq$. From the given relationship, $v = 50 - q/100$. Hence

$$a = 50; \qquad b = \frac{1}{100}$$

From equation (27.3), $p = c + d/v$. From the given relationship, $p = 1.25(5 + 300/v)$. Hence

$$c = 6.25; \qquad d = 375$$

From equation 27.4

$$\text{marginal cost} = c + \frac{d}{v} \times \frac{a}{v}$$

$$= 6.25 + \frac{18750}{v^2}$$

Flows, speeds, private and marginal costs are tabulated in table 27.3.

TABLE 27.3

Flow (p.c.e./h)	Speed $50 - q/100$ (km/h)	Private cost $1.25(5 + 300/v)$ (p/p.c.u. km)	Marginal cost $6.25 + 18750/v^2$ (p/p.c.u. km)	Demand $4 \times 10^4 \times p^{-1}$ (p.c.u./h)
200	48	14.06	14.39	2844
500	45	14.59	15.51	2742
1000	40	15.63	17.34	2560
1500	35	16.96	21.56	2360
2000	30	18.75	27.08	2134
2500	25	21.25	36.25	1882
3000	20	25.00	53.12	1600
4000	10	43.75	193.75	914

The private cost, marginal cost and demand curves are plotted and the intersection of the private cost and demand curves gives (a) the flow on the highway when each tripmaker is only aware of his private costs (see figure 27.10). The intersection of the marginal cost and demand curves gives (b) the flow on the highway when a trip is only made if the benefit of a trip to the tripmaker exceeds the additional cost imposed on other tripmakers. The ordinate XY gives the pricing charge that will result in flow condition (b). With the flows known the travel speeds can be calculated from equation 27.2, and are given below.

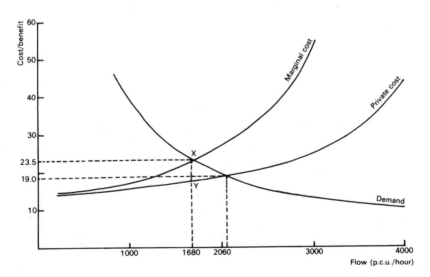

Figure 27.10

From the graphical plot:

flow condition (a)

$q = 2060$ p.c.u./h

From equation 27.2

$v = 29.40$ km/h

Flow condition (b)

$q = 1700$ p.c.u./h

From equation 27.2

$v = 33.00$ km/h

(c)   the pricing charge

$XY = 5.9$ p/km

PART 3[†]
TRAFFIC SIGNAL CONTROL

---

†See the Appendix, page 370, for a definition of the symbols used in part 3.

28

Introduction to traffic signals

It has been stated that the first traffic signal to be installed in Great Britain was erected in Westminster in 1868. It was illuminated by town gas and unfortunately for the future development of signals of this type was demolished by an explosion. Not until 1918 were signals used again for the control of highway traffic when manually operated three-colour light signals were introduced in New York. Some seven years later manually controlled signals were used in Piccadilly, followed in 1926 by the first automatic traffic signals in Great Britain, which were erected at Wolverhampton.

In the first signals alternate red and green fixed time periods were automatically timed and while they replaced police manpower they were not as efficient as manual control because they could not respond to changes in the traffic flow. As a refinement controllers were introduced, which were able to vary the relative durations of the fixed time green and red periods according to a preset timing pattern. It was thus possible to operate the signals with different sequences for the morning, mid-day and evening peak periods.

Where a major highway had several intersections along its length each controlled by traffic signals then means of allowing a nearly continuous progression of traffic along the major route were developed by linking the signals using a master timing device or controller instead of individual timing devices at each intersection.

During the early 1930s there was a further attempt to increase the ability of the signal controllers to deal with varying traffic demand by the incorporation of systems that would allow the signals to respond to individual vehicles. In some situations drivers were requested to sound their horns into microphones at the sides of the highway and later electrical contacts were operated by the passage of vehicles.

The vehicle detection method that became widely adopted in this period was the pneumatic tube detector and this survived in various forms in Great Britain until the 1960s. It had the disadvantages of being easily damaged by vehicles, particularly during snow-ploughing operations, and as these detectors were set in a concrete base obstructions to traffic were caused during installation and maintenance. They have now been replaced by the inductance detector, a cable set into the road surface that detects the passage or presence of a vehicle by a change in the electric field. The standard method of detection in the United Kingdom is known as 'System D' where detection for a green signal indication is carried out by a buried loop position-ed 40 metres from the stop line. Other loops to extend the existing green indication are sited between the initial loop and the stop line at distances of 12 metres and 25 metres from the stop line.

More recently, micro-wave detectors commonly seen on temporary traffic signals at roadworks have been introduced for permanent traffic signal installations. They offer the opportunity for lower installation costs and less traffic delay during any subsequent maintenance.

Problem

Which of the following signal timing devices or controllers would operate most efficiently when the traffic flows on the intersection-approach highways vary considerably:
(a) an isolated intersection where the signals are operated by a fixed-time controller;
(b) an intersection where the signals are operated by a fixed-time controller, which also controls adjacent intersections so as to assist progression along the major route;
(c) an intersection where the signals operate by vehicle actuation using inductance loops beneath the road surface?

Solution

(a) Incorrect. At isolated intersections where the signals are operated by a fixed-time controller the durations of the red and green periods are fixed with respect to average traffic flows. They are for this reason insensitive to short-term fluctuations in the traffic flow pattern.

(b) Incorrect. At an intersection where the signals are operated by a fixed-time controller, which also controls adjacent intersections so as to assist progression along the major route, major road vehicles are not unduly delayed when the traffic flows are similar to those for which the durations of the green and red periods were calculated. When the traffic flows fluctuate on a short-term basis these fixed red and green periods cause excessive delays to both major and minor road vehicles.

(c) Correct. At an intersection where the signals are actuated by vehicles passing over pneumatic or inductance detectors the signals respond to the vehicles on the traffic signal approaches and so are able to control traffic efficiently even when there are short-term variations in the flow.

29

Warrants for the use of traffic signals

A decision to use signal control in urban areas in preference to roundabout control or as a means of increasing traffic capacity at priority intersections may be made from the overall viewpoint of traffic management when urban computer control is being implemented.

A more detailed decision on the installation of traffic signals may also be made on the basis of traffic flow, pedestrian safety, accident experience and the elimination of traffic conflicts. Changes in these traffic characteristics consequent on the installation of signals can be determined and evaluated by cost–benefit methods.

Frequently the decision to install signal control is made by consideration of many issues, not all of which can be evaluated in traffic or monetary terms; some of these will be considered in greater detail.

When signal control is employed in preference to priority or roundabout control, then major road vehicles which previously passed unimpeded through the junction are on occasions required to stop and wait. On the other hand, minor road vehicles suffer reduced delay. Vehicular delay to the minor road stream under priority control can be estimated as described in chapter 18. Delay at roundabouts can be estimated using the ARCADY program developed by the Transport and Road Research Laboratory (see chapter 20). The delay to all traffic streams under signal control can be estimated as described in chapter 43.

Accident prediction at priority junctions and roundabouts can be predicted using the methods described in chapters 16 and 20 respectively. Accidents at signal controlled intersections can be estimated using the method employed in the COBA program.

The accident prediction formulae for traffic signal controlled intersections take the form

$$\text{Annual accidents} = x(f)^y$$

where x and y are constants which depend upon the form of the intersection and f is the product obtained by multiplying the combined inflow of vehicles from the two major opposing links by the sum of the inflows on the other minor links or, alternatively, f is the sum of the total inflow of vehicles into the junction. The first type of formula is referred to as a cross-product (C) model and the second as the inflow (I) model. Vehicles are expressed as thousands of vehicles per annual average day.

The type of formula used depends on the number of arms of the controlled junction and whether the highest standard of link at the intersection has a single or dual carriageway. Details are given in table 29.1.

TABLE 29.1   Formula type for accident prediction at signal-
controlled junctions

Junction reference number	No. of arms	Highest link standard	Formula type
1	3	single	I
2	3	dual	C
3	4	single	C
4	4	dual	C
5	5	single	I
6	5	dual	I

The values of the parameters x and y differ with junction form and also according to whether the junction is in an urban or rural situation. Values are given in table 29.2.

TABLE 29.2   Parameters for accident prediction at signal-
controlled junctions

Junction reference number	Rural		Urban	
	x	y	x	y
1	0.223	0.610	0.223	0.610
2	0.494	0.420	0.291	0.510
3	1.378	0.200	1.378	0.200
4	0.494	0.420	0.291	0.510
5	0.254	0.620	0.254	0.620
6	0.238	0.850	0.160	0.970

Accident experience can be worsened if a relatively small number of vehicles on the minor road experience difficulty in entering or crossing a nearly continuous major road traffic stream. In such circumstances the delays caused by the installation of signals are likely to be greater than under the previous priority control. Whilst this type of traffic conflict requires professional judgement, guidance is given for use in the United States. Here for 2 or more lane approaches on both major and minor roads, flows of 900 vehicles per hour, total both major road approaches, and 100 vehicles per hour on the most heavily trafficked minor road signals are justified when the flows exist for eight hours of an average day. Other values are given for different approach widths.

Signal control, even when pedestrian facilities are not provided, offers considerable assistance to pedestrian movement. In the United Kingdom the Department of Transport[1] advise that a separate pedestrian stage or one combined with a traffic stage may be required when the flow of pedestrians across any one arm of the junction is of the order of 300 per hour or more or when the turning traffic flowing into any arm has an average headway of less than 5 seconds during the time that such traffic can flow and is conflicting with a pedestrian flow of at least 50 pedestrians per hour. These traffic flows to be taken as the average of the 4 busiest hours over any weekday.

If a pedestrian facility is to be provided it may take the form of a full pedestrian stage where all traffic is stopped when pedestrians are allowed to cross all the arms of the junction. The pedestrian stage is demanded by push button but has the disadvantage of imposing additional delay on vehicular traffic.

A more efficient form of control from the viewpoint of vehicular movement is the use of a parallel pedestrian facility. This is achieved by banning some vehicular turning movements, but to prevent the danger which could be caused by illegal movements particular attention should be given to the kerb layout.

Where road layout permits, a large central island can be provided instead of the usual refuge so that pedestrians negotiate the road in two stages. The Department of Transport[1] recommend a minimum island size of 10 by 2.5 m.

As the provision of any pedestrian facility will normally reduce the proportion of green time available for vehicle movements, it may sometimes be necessary when a junction is close to capacity to install a pedestrian facility away from the junction. It is recommended that the crossing be not more than 50 m from the mouth of the junction; the pedestrian stage is incorporated within the junction signal cycle and the position of the stage chosen to minimise delay to traffic flow.

Reference

1. Department of Transport, Pedestrian facilities at traffic signal installations, *Advice Note* TA/15/81, London (1981)

Problem

Delays at a hypothetical T-junction under priority and signal control are illustrated below. Type A and type B priority junctions have good and poor visibility distances respectively for minor road vehicles entering the major road. Which of the following traffic situations would be likely to warrant a change from priority to signal control?

(a) An intersection where the priority control flow/delay curve is represented by curve B of figure 29.1, where the traffic signal flow/delay curve is as shown, and where the minor road flow is 200 veh/h.

(b) As above, but where the delay to minor road vehicles is 10 s and traffic volumes are not expected to increase rapidly in the future.

(c) As above, but where the priority control flow/delay relationship is illustrated by curve A, the major road flow is 800 veh/h, and the intersection is the scene of frequent traffic accidents.

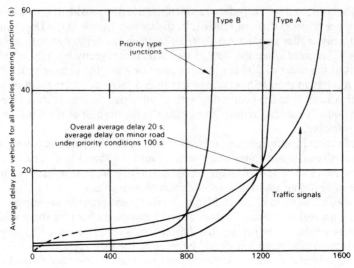

Solution

(a) Very likely. The minor road flow is 200 veh/h, and since the ratio of major/minor road flow is 4:1 the total flow entering the intersection is 1000 veh/h. From figure 29.1 the delay with traffic signal control is very considerably less than with priority control.

(b) Not likely. The delay to minor road vehicles is 10 s and since the delay to major road vehicles under priority control is zero the average delay to all vehicles entering the intersection is 2 s, which occurs when the total flow entering the intersection is approximately 400 veh/h. At this volume the delay at traffic signals is approximately 5 s.

(c) Likely. The major road flow is 800 veh/h and since this represents 0.8 of the total flow entering the intersection the delay with priority control for conditions of good visibility is less than with traffic signal control. The intersection is, however, the scene of frequent accidents and the increase in average delay due to the change to traffic signal control would probably be compensated by the decrease in accidents. As the traffic volumes are increasing rapidly, the change to signal control would be justified on delay considerations after a short period of time.

30

Phasing

In the control of traffic at intersections the conflicts between streams of vehicles are prevented by a separation in time. The procedure by which the streams are separated is known as phasing. A phase has been defined as the sequence of conditions applied to one or more streams of traffic, which during the cycle receive simultaneous identical signal indications.

The selection and use of phases is conveniently illustrated by the conventional cross roads where the major conflicts are between the north–south traffic stream and the east–west traffic stream. Because there are two major traffic conflicts then they may be resolved by two phases. The traffic movements during each of the phases are illustrated by figure 30.1.

In some intersection traffic situations there are more than two major traffic conflicts and then it is necessary to employ more than two phases in the traffic-control system. A typical situation is where at a normal cross roads there is a heavy right-turning movement on one of the approaches. There are now three major traffic conflicts and they may be resolved by the use of a three-phase control system. The traffic movements in this instance are illustrated by figure 30.2.

The number of phases employed at any intersection is kept to a minimum compatible with safety. The reason for this is explained subsequently in chapter 31.

There are two alternative ways in which the control of traffic movements at signals can be described: Phase control which refers to the periods of green and red time allocated to each traffic stream, and Stage control which refers to the sequential steps in which the junction control is varied. In figure 30.1 phase A refers to the signal indications that control the N/S traffic stream and associated turning movements whilst phase B refers in a similar manner to the E/W traffic stream. In figure 30.2 phases A, B and C refer to the signal indications, which control the N/S, W and E traffic streams respectively. In figure 30.3 the sequence of green/red signal indications control traffic on the E/W road is referred to as phase A whilst similarly phase B refers to the signal indications for the S approach.

On the other hand a stage usually commences from the start of an amber period and always ends at the start of the following stage. A stage usually, but not always, contains a green period and stages are arranged to follow each other in a predetermined order, although they may be omitted to reduce delay. In figure 30.1 stage 1 would be the green indication for the N/S traffic movements and it would be followed by stage 2 which would be the green indication for E/W traffic movements.

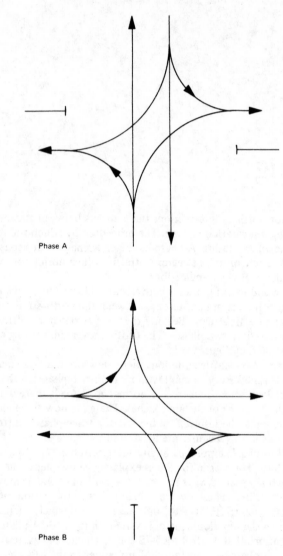

Figure 30.1   Traffic movements in a two-phase system

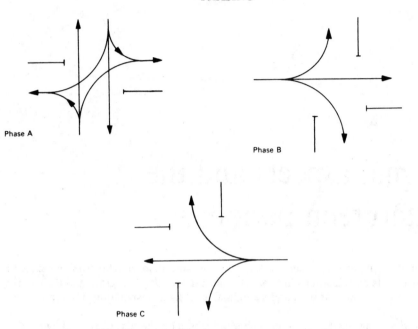

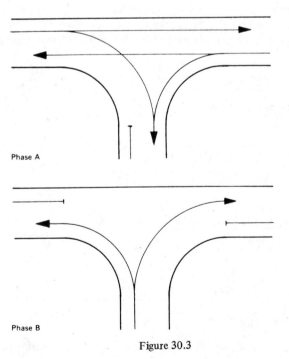

Figure 30.2 Traffic movements in a three-phase system

Figure 30.3

31

Signal aspects and the intergreen period

The indication given by a signal is known as the signal aspect. The usual sequence of signal aspects or indications in Great Britain is red, red/amber, green and amber. The amber period is standardised at 3 s and in all new signal installations the red/amber at 2 s.

In some older installations the amber indication on the first phase is shown concurrently with the red/amber indication on the second phase: in this case the red/amber indication has a duration of 3 s.

The period between one phase losing right of way and the next phase gaining right of way, that is the period between the termination of green on one phase and the commencement of green on the next phase, is known as the intergreen period.

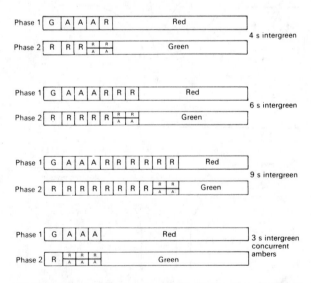

Figure 31.1   Examples of intergreen periods at a two-phase traffic signal

With modern controllers the minimum intergreen period is 4 s but this may be exceeded in particular circumstances. When the distance across the intersection is excessive the length of intergreen period must be based on the time required for a vehicle which passes over the stop line at the start of the amber period to clear a potential collision point with a vehicle starting at the onset of green of the following stage and travelling at the normal speed for the intersection. The Department of Transport[1] recommend intergreen periods based on this distance ranging from 5 to 12 seconds for distances of 9 to 74 metres for straight ahead movements. For turning movements the corresponding distances are 9 to 50 metres. The recommended values may have to be modified when a pedestrian stage is losing right of way, when the following stage may have to be delayed until the pedestrian area is clear.

When signals are located on higher speed roads a longer intergreen period provides a margin of safety for vehicles which are unable to stop on the termination of green. Controllers on high speed roads have the ability to extend the intergreen period on a maximum green termination.

Intergreen periods provide a convenient time during which right-turning vehicles can complete their turning movement after waiting in the centre of the intersection. Whilst the number of vehicles turning in this way is limited, it is useful when a separate right turning phase is not justified.

Referring to figure 31.1 it can be seen that within an intergreen period there is a period of time when all vehicle movement is prohibited, because the signal indication is either red or red/amber, equal to the intergreen period minus the 3 second amber period. For this reason an increase in the length of the intergreen period will result in a loss of traffic capacity and intergreen periods should have a minimum value consistent with safety.

The period of time lost to traffic flow is referred to as 'lost time' in the intergreen period and should not be confused with start and end lost times associated with green periods and dealt with subsequently in chapter 35.

Reference

1. Department of Transport, General principles of control by traffic signals, *Advice Note* TA/16/81, London (1981)

Problems

(a) State the factors which are considered in the selection of an appropriate intergreen period.
(b) When the intergreen period is 7 s what is the time between the last possible vehicle crossing the stop line on the approach losing right of way and the first possible crossing the stop line on the approach gaining right of way?

Solutions

(a) Factors considered in determining intergreen periods are:

(i)   the lengths of the travel paths within the intersection between the stop lines and potential collision points;

(ii)   the speeds of vehicles on the approaches and dangers which may exist on a maximum green termination;

(iii)   the time required for pedestrians to cross the carriageway when a pedestrian stage terminates;

(iv)   the clearance of right-turning vehicles waiting within the intersection.

(b) A vehicle may cross the stop line at amber if it cannot stop with safety and hence the period between a vehicle crossing the stop line on the approach losing right of way and a vehicle crossing the stop line on an approach gaining the right of way is $7 - 3 = 4$ s.

32
Signal control strategies

The fixed time operation of signals where the stage changes and the lengths of the green and red periods are fixed throughout the day is usually an unsatisfactory method of control, producing frustration to drivers during low flow periods when they are held at a red signal without any conflicting traffic being apparent.

Control strategy is usually flexible and is achieved by one of the following methods:

(a) by vehicle actuation;
(b) by cableless linking of signals;
(c) by a cable linked system;
(d) by an integral time switch.

If vehicle actuation is employed then a series of buried loops are placed on the approaches with the initial detector some 40 metres distant from the stop line. When this method of control is used then the method of control is varied in the following manner.

When a green signal is displayed it is desirable for the green indication to be shown for an initial fixed period which cannot be overriden by other demands. Such a period is built into traffic signal controllers, its value normally being 7 seconds.

Exceptions are when pedestrians walk with green on a stage that does not allow vehicles to cross their path, when the minimum green must be related to the pedestrian crossing time. When the proportion of heavy vehicles on an approach is high and they accelerate slowly owing to the gradient, then the minimum green time may be increased. On late start and early cut-off stages shorter minimum green times are frequently used.

A vehicle detected on an approach during the display of the green indication will normally extend the period of green so that a vehicle can cross the stop line before the expiry of green. Usually there are three loops on an approach, each one of which extends the green time by 1.5 seconds. If the approach has a steep gradient a longer extension period will be required.

When a vehicle is detected approaching a red signal indication the demand for the green signal is stored in the controller which serves stages in cyclic order and omits any stages for which a demand has not been received.

The demand for the green stage is satisfied when the previous stage that showed a green indication has exceeded its minimum green period and there has not been a demand for a green extension on the running stage or the last vehicle extension on the running stage has elapsed and there has not been a further demand. Alternatively, the demand for the green stage is satisfied if, after the demand is entered in the controller, the running stage runs for a further period of time known as the maximum green time. This would occur if there were continuous demands for green on the running stage. The first type of change is known as a gap change and is safer than the second type which is known as a maximum green change.

Calculation of the maximum green time depends on traffic flow conditions and is described subsequently.

When traffic flow changes during the day can be predicted with some certainty then cableless linking can offer benefits to traffic movement. Co-ordination between signals at adjacent intersections can be achieved by the use of controllers which are synchronised by the mains supply frquency and which also incorporate a solid state memory store containing stage timings, cycle times and offset periods. These plans are selected by time of day, or day of the week, according to changes in the traffic flow. It is also possible to incorporate one demand-dependent stage selected either by vehicle detection or by a pedestrian demand.

An older system which is being superseded by cableless linking, is cable linking. Here information is passed between controllers so that the commencement of a selected stage at a key intersection controls the beginning or end of other selected stages at other intersections. It is usually necessary in linked systems such as these for a common cycle time to be imposed so that co-ordination with the key intersection may be maintained.

Traffic signal control over a wide area can also be controlled by instructions from a central computer. Cycle time, the start and the length of the green stage together with the offsets between green stages at adjacent controllers are controlled by a central computer using plans determined by historic records of traffic flow or reacting to the changes in detected traffic flows in a dynamic manner. The subject of computer control of signals on an area wide basis is considered in detail in chapter 51.

Problem

For the effects a, b, c, d given, select the appropriate causes.

Effects:

(a) The red signal indication on a traffic-signal approach changes to red/amber.
(b) The green signal indication on a traffic-signal approach changes to amber.
(c) The green signal indication on a traffic-signal approach remains at green.
(d) The red signal indication on a traffic-signal approach remains at red.

Causes:

(e) The green indication on the preceding phase has run to maximum green and there has been a demand for green on the approach being considered.
(f) There has been a demand for the extension of the green period on the preceding phase and that phase has not yet reached its maximum green period.

(g) The minimum green period on the approach being considered has not as yet expired.

(h) The last vehicle extension on the phase being considered has expired and there has been a demand for the green signal on the next phase.

Solution

The effect (a) is the result of cause (e).
The effect of (b) is the result of cause (h).
The effect (c) is the result of cause (g).
The effect (d) is the result of cause (f).

33

Geometric factors affecting the capacity of a traffic signal approach

The capacity of a signal-controlled intersection is limited by the capacities of individual approaches to the intersection. There are two types of factors which affect the capacity of an approach: geometric factors which are considered in this chapter, and traffic and control factors which are discussed subsequently.

The capacity of a traffic signal approach is the sum of the capacities or saturation flows of the individual lanes comprising the approach. The saturation flow is independent of traffic and control factors and is the maximum flow, expressed in equivalent passenger car units, that can be discharged from a traffic lane when there is a continuous green indication and a continuous queue on the approach.

Geometric factors affecting lane saturation flow are: the position of the lane (nearside or non-nearside), the width of the lane and its gradient, and the radius of any turning movements. Recent research[1] has indicated the following values as being appropriate for the United Kingdom.

For unopposed streams in individual traffic lanes, the saturation flow S_1 is given by

$$S_1 = (S_0 - 140d_n)/(1 + 1.5f/r) \text{ pcu/h}$$

where $S_0 = 2080 - 42 d_g \times G + 100 (w - 3.25)$

and d_n is 1 for nearside lanes and 0 for non-nearside lanes
d_g is 1 for uphill entries and 0 for downhill entries
G is the percentage gradient for the entry
w is the lane width at entry (m)
f is the proportion of turning vehicles in the lane
r is the radius of curvature of vehicle paths (m)

For opposed streams containing opposed right-turning traffic in individual lanes the saturation flow S_2 is given by

$$S_2 = S_g + S_c$$

where S_g is the saturation flow in lanes of opposed mixed turning traffic during the effective green period (pcu/h)

S_c is the saturation flow in lanes of opposed mixed turning traffic after the effective green period (veh/h or pcu/h)

$$S_g = (S_0 - 230)/(1 + (T - 1)f)$$

and $\quad T = 1 + 1.5/r + t_1/t_2$

$$t_1 = 12X_0^2/(1 + 0.6(1 - f)N_s)$$

$$t_2 = 1 - (fX_0)^2$$

$$S_c = P(1 + N_s)(fX_0)^{0.2}\, 3600/\lambda c$$

$$X_0 = q_0/\lambda n_\varrho\, s_0$$

where X_0 is the degree of saturation on the opposing arm, that is, the ratio of the flow on the opposing arm to the saturation flow on that arm

N_s is the number of storage spaces available inside the intersection which right turners can use without blocking following straight ahead vehicles

λ is the proportion of the cycle time effectively green for the phase being considered, that is, the effective green time divided by the cycle time

c is the cycle time (s)

q_0 is the flow on the opposite arm expressed as vehicles per hour of green time and excluding non-hooking right turners

p is a conversion factor from vehicles to pcu and is expressed as
$$1 + \sum_i (\alpha_i - 1)p_i$$

where α_i is the pcu value of vehicle type i

p_i is the proportion of vehicles of type i in the stream

n_ϱ is the number of lanes on the opposing entry

s_0 is the saturation flow per lane for the opposite entry (pcu/h)

T is the through car unit of a turning vehicle in a lane of mixed turning traffic, each turning vehicle being equivalent to T straight ahead vehicles

Most traffic signal approaches are marked out in several lanes and the total saturation flow for the approach is then the sum of the saturation flows of the individual lanes.

Reference

1. R. M. Kimber, M. McDonald and N. B. Hounsell, The prediction of saturation flows for road junctions controlled by traffic signals, *Transport and Road Research Laboratory Research Report* 67, Crowthorne (1986)

Problems

1. A nearside unopposed lane of a traffic signal approach has a width at entry of 2.4 m and an uphill gradient of 5 per cent; 25 per cent of vehicles turn left with a turning radius of 20 m. Calculate the saturation flow for this lane.

2. A non-nearside lane of a traffic signal approach has a width at entry of 3.0 m and a downhill gradient of 3 per cent; 40 per cent of vehicles turn right with a turning radius of 25 m. The cycle time is 60 s and the effective greentime is 40 s. The right-turning vehicles are opposed by a straight ahead lane with a degree of saturation of 0.85. Two right-turning vehicles may wait within the intersection without obstruction to following straight ahead vehicles and the ratio of pcu/vehicles is 1.5. Calculate the saturation flow for this lane.

Solutions

1. For a non-opposed lane the saturation flow is calculated from

$$S_1 = (S_0 - 140\,d_n)/(1 + 1.5\,f/r) \text{ pcu/h}$$

where $S_0 = 2080 - 42\,d_g \times G + 100(w - 3.25)$

$$= 2080 - 42 \times 1 \times 5 + 100(2.4 - 3.25)$$

$$= 1770.85$$

$$S_1 = (1770.85 - 140 \times 1)/(1 + 1.5 \times 0 \times 25/20)$$

$$= 1601 \text{ pcu/h}$$

2. For an opposed lane the saturation flow is calculated from

$$S_2 = S_g + S_c$$

$$S_g = (S_0 - 230)/(1 + (T - 1)f)$$

and $T = 1 + 1.5/r + t_1/t_2$

$$t_1 = 12X_0^2/(1 + 0.6(1 - f)N_s)$$

$$t_2 = 1 - (fX_0)^2$$

$$t_1 = 12 \times 0.85^2/(1 + 0.6(1 - 0.4)2)$$

$$= 5.04$$

$$t_2 = 1 - (0.4 \times 0.85)^2$$

$$= 0.88$$

$$T = 1 + 1.5/25 + 5.04/0.88$$

$$= 6.79$$

$$S_0 = 2080 - 42\,d_g \times G + 100(w - 3.25)$$

$$= 2080 - 42 \times 0 \times 3 + 100(3.0 - 3.25)$$

$$= 2055$$

$$S_g = (2055 - 230)/(1 + (6.79 - 1)0.40)$$

$$= 550 \text{ pcu/h}$$

$$S_c = P(1 + N_s)(fX_0)^{0.2}\, 3600/\lambda c$$

$$= 1.5\,(1 + 2)\,(0.4 \times 0.85)^{0.2}\, 3600/40$$

$$= 326 \text{ pcu/h}$$

$$S_2 = S_g + S_c$$

$$= 550 + 326 \text{ pcu/h}$$

$$= 876 \text{ pcu/h}$$

34

The effect of traffic factors on the capacity of a traffic signal approach

The effect of traffic factors on the capacity of a traffic signal approach is usually allowed for by the use of weighting factors, referred to as 'passenger car units', assigned to differing vehicle categories. As a consequence the saturation flow of a signal approach or of a single approach lane is expressed in passenger car units per hour (pcu/h).

Whilst the effect of vehicle type on capacity might depend upon the geometric design of the junction, that is, large vehicles might be affected by turning radius or heavy vehicles by gradient, it has been usual in the United Kingdom to use constant factors independent of junction geometry.

The passenger car unit values recommended for use in United Kingdom junction design, with the exception of those for two-wheel vehicles, have been determined using observations of headway ratios[1].

In this method the time headways between individual vehicles crossing the signal approach stop line are measured and time headways between individual pairs of vehicles of interest are grouped. For instance, if it was required to determine the passenger car equivalent of a medium commercial vehicle (defined as a vehicle with 2 axles but more than 4 wheels) then pairs of headways between these vehicles and light vehicles (defined as 3 or 4 wheel vehicles) would be grouped as follows.

1. A light vehicle following a light vehicle.
2. A light vehicle following a medium commercial vehicle.
3. A medium commercial vehicle following a light vehicle.
4. A medium commercial vehicle following a medium commercial vehicle.

Vehicles crossing the stop line within 3 seconds of the commencement and termination of the green period should not be included in the observations because of the effects of acceleration and deceleration.

The passenger car unit for a medium commercial vehicle can be found by dividing the mean headway for a medium goods vehicle following a medium goods vehicle by the mean headway for a light vehicle following a light vehicle. This is true if the effect of a medium commercial vehicle is independent of whether the vehicles preceding it and following it are light or medium commercial vehicles. The necessary and sufficient condition for this is that the sum of the average headways for a light vehicle following a light vehicle and a medium commercial vehicle following a medium commercial vehicle should be equal to the sum of the average headways for a light vehicle following a medium commercial vehicle and a medium commercial vehicle following a light vehicle.

If this condition is not met then corrected values of the mean headways can be calculated as follows.

Corrected value of mean headway for a light vehicle following a light vehicle equals the uncorrected value (w) minus the correction factor divided by the number of headways in this group (a).

Corrected value of mean headway for a light vehicle following a medium commercial vehicle equals the uncorrected value (x) plus the correction factor divided by the number of headways in this group (b).

Corrected value of mean headway for a medium commercial vehicle following a light vehicle equals the uncorrected value (y) plus the correction factor divided by the number of headways in this group (c).

Corrected value of mean headway for a medium commercial vehicle following a medium commercial vehicle equals the uncorrected value (z) minus the correction factor divided by the number of headways in this group (d).

The correction factor is given by

$$\frac{abcd\,(w - x - y + z)}{bcd + acd + abd + abc}$$

As a result of investigations carried out by Martin and Voorhees Associates, Southampton University and the Transport and Road Research Laboratory, values of passenger car equivalents have been proposed for use in United Kingdom signal design[2] and are given below.

Light vehicles (3 or 4 wheeled vehicles)	1.0
Medium commercial vehicles (2 axles but more than 4 wheels)	1.5
Heavy commercial vehicles (vehicles with more than 2 axles)	2.3
Buses and coaches	2.0
Motorcycles	0.4
Pedal cycles	0.2

The number of vehicles crossing the stop line in a given period of time depends not only on the compaction of the traffic and the saturation flow, but also on the proportion of time during which the signal is effectively green (λ).

A cycle is a complete sequence of signal indications, that is, a green period and a red period for a two-phase system, and the time during which the signal is effectively green during a cycle, known as the 'effective green time'. The maximum number of vehicles crossing the stop line per hour is then

$$\frac{\text{saturation flow} \times \text{effective green time}}{\text{cycle time}}$$

$$= \text{saturation flow} \times \lambda$$

where the saturation flow is in vehicles per hour. It is however usual to express saturation flow in terms of passenger car units per hour, and the flow across the stop line during one hour is then expressed in passenger car units.

References

1. D. A. Scraggs, The passenger car equivalent of a heavy vehicle in single lane flow at traffic signals, *Road Research Laboratory Report* LN/573/DAS, Transport and Road Research Laboratory, Crowthorne (1964)
2. R. M. Kimber, M. McDonald and N. B. Hounsell, The prediction of saturation flows for road junctions controlled by traffic signals, *Research Report 67*, Transport and Road Research Laboratory, Crowthorne (1986)

Problems

1. The hourly traffic flow on an unopposed non-nearside 3.8 m level lane of an approach where turning vehicles follow a path with a radius of curvature of 20 m is 400 passenger cars, 100 medium goods vehicles and 40 motor cycles; one-fifth of all vehicles turn. If the approach is just able to pass all the traffic, should the ratio of effective green time to cycle time be 0.27, 0.33 or 0.55?

2. The following headways were obtained on a traffic signal approach. Calculate the passenger car equivalent of a medium commercial vehicle.

	Light vehicle following light vehicle	Light vehicle following medium commercial	Medium commercial following light vehicle	Medium commercial following medium commercial
Number of headways	66 (a)	33 (b)	32 (c)	9 (d)
Mean headway (seconds)	1.9 (w)	2.8 (x)	2.3 (y)	3.0 (z)

Solutions

1. Converting the flow to passenger car units

400 passenger cars	$= 400 \times 1$	$= 400$ pcu
100 medium goods vehicles	$= 100 \times 1.5$	$= 150$ pcu
40 motor cycles	$= 40 \times 0.4$	$= \underline{16}$ pcu
		566 pcu

Saturation flow $= (S_0 - 140 \, d_n)/(1 + 1.5 \, f/r)$

$$S_0 = 2080 - 42 \, d_g \, G + 100(w_\varrho - 3.25)$$

$$= 2080 + 100(3.8 - 3.25)$$

$$= 2135$$

Saturation flow $= 2135/(1 + 1.5 \times 0.20/20)$

$$= 2103 \text{ pcu/h}$$

Maximum flow over the stop line $= \dfrac{\text{saturation flow} \times \text{effective green}}{\text{cycle time}}$

$$566 = \dfrac{2103 \times \text{effective green}}{\text{cycle time}}$$

$$\dfrac{\text{effective green}}{\text{cycle time}} = \dfrac{566}{2103}$$

$$= 0.27$$

2. An initial check must be made to determine if a correction is necessary. This is not required if 1.9 + 3.0 equals 2.8 + 2.3; this is not so and hence correction is therefore necessary.

Correction factor $= \dfrac{abcd\,(w - x - y + z)}{bcd + acd + abd + abc}$

$$= \dfrac{66 \times 33 \times 32 \times 9\,(1.9 - 2.8 - 2.3 + 3.0)}{33 \times 32 \times 9 + 66 \times 32 \times 9 + 66 \times 33 \times 9 + 66 \times 33 \times 32}$$

$$= \dfrac{-125452.8}{117810}$$

$$= -1.0649$$

Corrected value of mean headway for a light vehicle following a light vehicle is given by

$$1.9 - (-1.0649/66)$$

$$= 1.9 \text{ s}$$

Corrected value of mean headway for a medium commercial vehicle following a medium commercial vehicle is given by

$$3.0 - (-1.0649/9)$$

$$= 3.1 \text{ s}$$

The passenger car equivalent of a medium commercial vehicle is given by 3.1/1.9 or 1.6.

35

Determination of the effective green time

In chapter 34 the concept of effective green time was introduced as a means of determining the number of vehicles that could cross a stop line over the whole of the cycle comprising both red and green periods.

It is obvious that in practice the flow across the stop line cannot commence or terminate instantly because at the end of the green indication this flow is slowly reduced to zero.

A study of the discharge of vehicles across the stop line allows the effective green time to be determined. Figure 35.1 shows the variation of discharge with time, the area beneath the curve representing the number of vehicles that cross the stop line during the green period. The area beneath the curve is not easily determined and for convenience a rectangle of equal area to that under the curve is superimposed upon the curve. The height of the rectangle is equal to the saturation flow and the base of the rectangle is the effective green time.

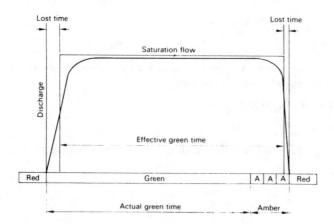

Figure 35.1 Variations in the discharge across the stop line

288

As the amber indication is a period during which, under certain circumstances, vehicles may cross the stop line the discharge across the stop line commences at the beginning of the green period and terminates at the end of the amber period. The time intervals between the commencement of green and the commencement of effective green and also between the termination of effective green and the termination of the amber period are referred to as the lost time due to starting delays.

From figure 35.1 it can be seen that the actual green time plus the amber period is equal to the effective green time plus the lost time due to starting delays.

In actual practice the lost time due to starting delays is taken as 2 s and so the effective green time is equal to the actual green time plus the 3 s amber period minus the 2 s lost time.

Determination of the lost time on a traffic-signal approach

The observations made in chapter 33 will be used to determine the lost time due to starting delays at a traffic signal-controlled intersection in the City of Bradford. The observations are repeated below in table 35.1.

TABLE 35.1

Time (minutes)	0	0.1	0.2	0.3	0.4	0.5
No. of vehicles crossing stop line	60	76	71	78	79	
No. of saturated intervals observed	32	32	32	32	32	
Discharge per 0.1 minute	1.88	2.48	2.22	2.44	2.47	

No. of last saturated intervals 24, average duration 5.9 s

Total duration of the last saturated intervals   = 142 seconds

Total number of vehicles crossing stop line   = 41

Discharge per 0.1 minute during last saturated interval = 41 × 6/142 = 1.74 vehicles

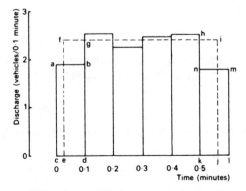

Figure 35.2   Observed discharge across the stop line

Additionally, it has been previously calculated in chapter 33 that the saturation flow was 1670 p.c.u./h or 2.40 vehicles/0.1 minute.

The lost time at the beginning and end of the green period may be calculated by reference to figure 35.2. By definition the number of vehicles represented by the rectangle efij is equal to the number of vehicles represented by the original histogram (because total flow during effective green time is equal to total flow during green plus amber period). The number of vehicles represented by the area dghk is also equal to the number of vehicles represented by the four 0.1 minute periods of saturated flow between d and k.

This means that:

1. The number of vehicles represented by abdc is equal to the number of vehicles represented by efgd.
2. The number of vehicles represented by hijk is equal to the number of vehicles represented by nmlk.

That is

$$ed \times 2.40 = 1.88 \times 0.1$$

giving

$$ed = 0.08 \text{ minute}$$

or

$$ce = 0.02 \text{ minute}$$

similarly

$$jl = 0.03 \text{ minute}$$

and

$$jl + ce = \text{lost time during green phase} = 2.9 \text{ s}$$

This value is higher than the accepted value of 2 s and the saturation flow was also noted to be lower than the usual design value. These departures from the accepted values were considered to be caused by a high proportion of elderly drivers in the traffic flow.

Problem

The lost time due to starting delays on a traffic signal approach is noted to be 3 s; the actual green time is 25 s. Is the effective green time 24 s, 25 s or 26 s?

Solution

$$\text{Effective green time} = 25 \text{ s} + 3 \text{ s (amber)} - 3 \text{ s (lost time)}$$

$$= 25 \text{ s}$$

36

Optimum cycle times for an intersection

The length of the cycle time under fixed time operation is dependent on traffic conditions. Where the intersection is heavily trafficked cycle times must be longer than when the intersection is lightly trafficked.

One definition of the degree of trafficking of an approach is given by the y value, which is the flow on the approach divided by the saturation flow.

For any given traffic-flow conditions with the signals operating under fixed-time control, the duration of the cycle must affect the average delay to vehicles passing through the intersection. Where the cycle time is very short, the proportion of the cycle time occupied by the lost time in the intergreen period and by starting delays is high, making the signal control inefficient and causing lengthy delays.

When on the other hand the cycle time is considerably longer, waiting vehicles will clear the stop line during the early part of the green period and the only vehicles crossing the stop line during the latter part of the green period will be those that subsequently arrive, often at extended headways. As the discharge rate or saturation flow across the stop line is greatest when there is a queue on the approach this also results in inefficient operation.

As a result of the computer simulation of flow at traffic signals carried out by the Road Research Laboratory[1], it was possible to show these variations of average delay with cycle time that occur at any given intersection when the flows on the approaches remain constant. They are illustrated in figure 36.1 and can be explained for most practical purposes by the total lost time and the y value found at the intersection.

It is shown in *Road Research Technical Paper* 39 that a sufficiently close approximation to the optimum cycle time C_0 could be obtained by the use of the following equation

$$C_0 = \frac{1.5L + 5}{1 - Y}$$

where L is the total lost time per cycle

Y is the sum of the maximum y values for all the phases comprising the cycle.

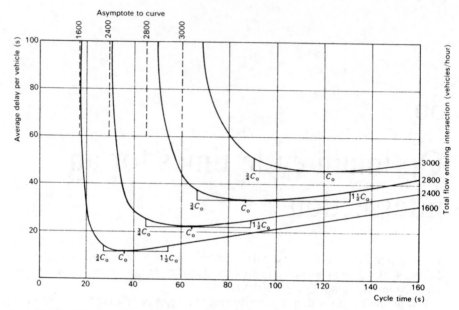

Figure 36.1 Effect on delay of variation of the cycle length (based on ref. 1)
2-phase, 4-arm intersection, equal flows on all arms, equal saturation flows of 1800 vehicles/
hour, equal green times, total lost time/cycle 10 s

The application of this formula may be illustrated from figure 36.1 and the calculation of the optimum cycle time C_0 is tabulated in table 36.1 for each of the four flow rates illustrated. An inspection of figure 36.1 shows that the values by the approximate formula are similar to those obtained by simulation.

TABLE 36.1

(1) Total flow (veh/h)	(2) Flow/approach (veh/h)	(3) y value (2)/1800	(4) L	(5) Y $(n \times (3)^*)$	(6) C_0 (s)
3000	750	0.42	10	0.84	125
2800	700	0.39	10	0.78	91
2400	600	0.33	10	0.66	59
1600	400	0.22	10	0.44	36

*n = number of phases.

There is a minimum cycle time of 25 s fixed from safety considerations while a maximum cycle time of 120 s is generally considered desirable. Calculated optimum cycle times should normally be limited to this range of values.

The calculation of the optimum cycle time consists of several steps, best illustrated by the flow chart given in figure 36.2.

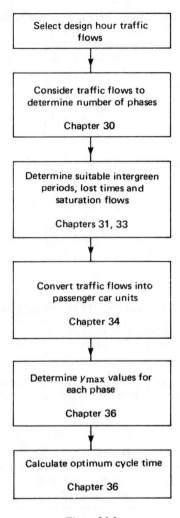

Figure 36.2

Problem: Optimum cycle times for an intersection

Design hour traffic flows at a 4-arm 3-phase intersection are given in table 36.2; left-turn vehicles comprise 20 per cent of the total lane flow. Two right-turn vehicles can wait within the intersection without delay to following vehicles. The intergreen period is 5 s, start and end lost times are 2 s per green period. All approaches are level and the radius of curvature of all turning paths is 20 m. Is the optimum cycle time 60 s, 90 s or 120 s?

TABLE 36.2 Design-hour traffic flows

	Light vehicles	Medium goods	Buses	Motor cycles	Approach width (m)
North approach, straight ahead and left turning	500	100	10	20	3.0
North approach, right turning	50	10	0	10	3.0
South approach, straight ahead and left turning	400	150	0	30	3.0
South approach, right turning	40	30	0	5	3.0
West approach, straight ahead and left turning	400	50	5	10	3.65
West approach, right turning	300	60	0	20	3.65
East approach, straight ahead and left turning	200	180	4	15	3.65
East approach, right turning	260	20	0	10	3.65

Solution

For the traffic flows given in table 36.2 the three major traffic movements are:

> north/south, all directions of movement;
> east/west, straight ahead and left-turning movements;
> east/west, right-turning movements.

These three major traffic movements will each be given a separate phase.

The traffic flow for each of these phases will now be converted to passenger car equivalents (table 36.3).

TABLE 36.3 Design hour traffic flow in pcus

	Light vehicles	Medium goods	Buses	Motor cycles	Total
North approach, ahead and left	500	150	20	8	678
North approach, right	50	15	0	4	69
South approach, ahead and left	400	225	0	12	637
South approach, right	40	45	0	2	87
West approach, ahead and left	400	75	10	4	489
West approach, right	300	90	0	8	398
East approach, ahead and left	200	270	8	6	484
East approach, right	260	30	0	4	294

The saturation flows of the approaches will now be calculated (see chapter 33).

For non-opposed flows.
North approach, ahead and left (nearside lane).

$S_0 = 2080 + 100 \, (3.0 - 3.25)$

$= 2055$

$S_1 = (2055 - 140)/(1 + 1.5 \times 0.2/20)$

$= 1887 \, \text{pcu/h}$

South approach, ahead and left (nearside lane).

$S_1 = 1887 \, \text{pcu/h}$

West approach, ahead and left (nearside lane).

$S_0 = 2080 + 100 \, (3.65 - 3.25)$

$= 2120$

$S_1 = (2120 - 140)/(1 + 1.5 \times 0.2/20)$

$= 1951 \, \text{pcu/h}$

East approach, ahead and left (nearside lane).

$S_1 = 1951 \, \text{pcu/h}$

West approach, right.

$S_0 = 2080 + 100 \, (3.65 - 3.25)$

$= 2120$

$S_1 = 2120/(1 + 1.5/20)$

$= 1972 \, \text{pcu/h}$

East approach, right.

$S_1 = 1972 \, \text{pcu/h}$

For opposed flows.

North approach, right and South approach, right.

An inspection of the right-turning flows and the opposing flows indicates that the south approach right-turn flow is the critical one to be used in design.

$S_2 = S_g + S_c$

$S_g = (S_0 - 230)/(1 + (T - 1)f)$

$S_0 = 2055$

$t_1 = 12X_0^2/(1 + 0.6(1 - f)N_s)$

assume $X_0 = 0.75$

$t_1 = 12 \times 0.75^2/(1 + 0.6(1 - 1)2)$

$= 6.75$

$t_2 = 1 - (fX_0)^2$

$$= 1 - (1 \times 0.75)^2$$

$$= 0.44$$

$$T = 1 + 1.5/r + t_1/t_2$$

$$= 1 + 1.5/20 + 6.75/0.44$$

$$= 16.42$$

$$S_g = (2055 - 230)/(1 + (16.42 - 1)1)$$

$$= 111 \text{ pcu/h}$$

$$S_c = P(1 + N_s) (fX_0)^{0.2} 3600/\lambda_c$$

$$P = 87/75 \text{ pcu/vehicle}$$

$$= 1.16 \text{ pcu/vehicle}$$

Assume λ_c the effective green time/cycle is 30 s.

$$S_c = 1.16(1 + 2) (1 \times 0.75)^{0.2} 3600/30$$

$$= 1.16 \times 3 \times 0.94 \times 120$$

$$= 393 \text{ pcu/h}$$

$$S_2 = 111 + 393 \text{ pcu/h}$$

$$= 504 \text{ pcu/h}$$

The flows, saturation flows and y values, the ratio of flow to saturation flow, for each approach are now tabulated in table 36.4.

TABLE 36.4　Flows, saturation flows and y values

	Flow	Saturation flow	y
North approach, ahead and left	678	1887	0.36
North approach, right	69	–	–
South approach. ahead and left	637	1887	0.34
South approach, right	87	504	0.17
West approach, ahead and left	489	1951	0.25
West approach, right	398	1972	0.20
East approach, ahead and left	484	1951	0.25
East approach, right	294	1972	0.15

Considering the three major traffic movements, the critical approaches to be used in design (the y_{max} values) are:

for the north/south movements the maximum value of y occurs on the north approach, ahead and left, and is equal to 0.36

for the east/west, straight ahead and left, the maximum value of y occurs on the west approach and is equal to 0.25

for the east/west right turning movements the maximum value of y occurs on the west approach and is equal to 0.20

The total last time per cycle (L) is composed of the lost time during the green period and the lost time during the intergreen period.

The lost time during the 5 s intergreen period is 2 s and the intergreen period occurs three times with a three-phase system (see chapter 31). Start and end lost times total 2 s for each of the three green periods of the three-phase cycle (see chapter 33). The total lost time (L) is $3 \times 2 + 3 \times 2 = 12$ s. The optimum cycle time C_0 is given by

$$C_0 = \frac{1.5L + 5}{1 - Y}$$

$Y = \Sigma y_{max} = 0.36 + 0.25 + 0.20 = 0.81$

$$C_0 = \frac{1.5 \times 12 + 5}{1 - 0.81} = 121 \text{ s}$$

It is usual to limit cycle times to a maximum value of 120 s and in this example this maximum value will be adopted.

37

The timing diagram

Previously that optimum cycle time has been calculated which would result in minimum overall delay when employed with fixed-time signals. The intergreen periods have also been selected previously and the remaining calculation, before the whole sequence of signal aspects can be described, is to calculate the duration of the green signal aspects.

Once the duration of these green aspects has been calculated, they can be employed with vehicle-actuated signals, and are then the maximum green times at the end of which a phase change will occur regardless of any demands for vehicle extensions.

The first step is to calculate the amount of effective green time available during each cycle from

available effective green time/cycle = cycle time − lost time/cycle (L)

This available effective green time is then divided between the phases in proportion to the y_{max} value for each phase.

For the particular value of lost time due to starting delays associated with each green period, the actual green time can be calculated. This is the indication shown on the signal and depicted on the timing diagram.

These steps will now be illustrated by the following example.

C_0 82 s
L 12 s
y_{max} north/south phase (all movements) 0.21
y_{max} east/west phase (straight ahead and left turn) 0.26
y_{max} east/west phase (right turning) 0.25

available effective green time/cycle = 82 − 12 = 70 s

This is divided between the phases in the ratio

0.21:   0.26:   0.25

or

20.0 s,   25.0 s   and   25.0 s

When the lost time due to starting delays is 2 s then the actual green time is equal to the effective green time minus 1 s (see chapter 35).

298

The actual green times used are in practice

    north/south traffic flow   19 s
    east/west straight ahead and left turning   24 s
    east/west right-turning flows   24 s

The actual green times have been rounded to the nearest second.
Figure 37.1 is a timing diagram showing these signal indications.

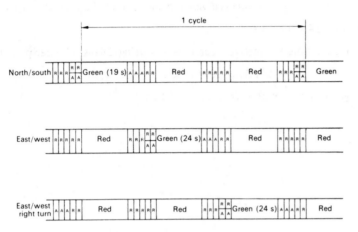

Figure 37.1   A three-phase timing diagram

Problems

An intersection is controlled by four-phase traffic signals, with a cycle time of 100 s.
The minimum intergreen period is employed and a value of lost time per green
period of 3 s is assumed. Saturation flows on all approaches are identical, but the
maximum traffic flows on two of the phases are twice the maximum traffic flows on
the remaining two phases. Which of the following series of actual green times would
be appropriate:

(a) 28 s, 28 s, 14 s and 14 s;
(b) 30 s, 15 s, 30 s and 15 s;
(c) 30 s, 15 s, 15 s and 15 s;
(d) 14 s, 14 s, 30 s and 30 s.

Solutions

The minimum intergreen period that can be employed is 4 s (see chapter 31), and
when it has this value the lost time in the intergreen period is 1 s.
    The total lost time L per cycle is equal to the number of phases multiplied by
the lost time in the intergreen period plus the lost time caused by starting delays,
that is $4(1 + 3) = 16$ s (see chapter 35).

The y value for an approach is the flow divided by the saturation flow, and the maximum flows for each phase are in the ratio: 2:2:1:1. As the saturation flows on each approach are equal, this is also the ratio of the y values.

available effective green time/cycle = cycle time − lost time/cycle

$$= 100 - 16 \text{ s}$$

$$= 84 \text{ s}$$

This available effective green time is divided in the ratio of the y values, that is

28 s, 28 s, 14 s and 14 s

actual green time = effective green time + starting delays − amber period

$$= \text{effective green time} + 3 \text{ s} - 3 \text{ s}$$

The actual green times are 28 s, 28 s, 14 s and 14 s.

38

Early cut-off and late-start facilities

Where the number of right-turning vehicles is not sufficient to justify the provision of a right-turning phase but where right-turning vehicles have difficulty in completing the traffic movement, then an early cut-off or a late start of the opposing phase is employed.

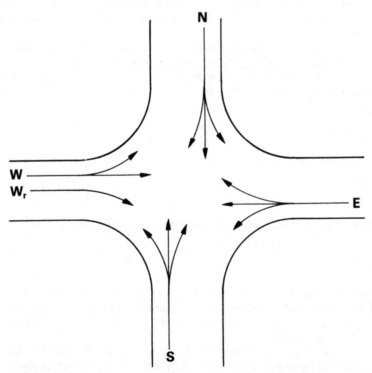

Figure 38.1   Traffic flows at an intersection with a heavy right-turning movement

An early cut-off of the opposing flow allows right-turning vehicles to complete their traffic movement at the end of the green period when the opposing flow is halted. To allow straight-ahead vehicles on the same approach to flow without interruption during the early part of the green period, it is necessary for there to be sufficient room in the intersection for right-turning vehicles to wait.

In contrast a late-start facility discharges the right-turning vehicles at the commencement of the green period and for this reason storage space is not as important as with the early cut-off facility.

The calculation of optimum cycle times for fixed-time operation and hence the determination of maximum green times for vehicle-actuated operation is as follows.

The traffic flows are as shown in figure 38.1 and it is necessary to introduce a late-start or early cut-off facility to permit vehicles to turn right from the west approach.

These traffic flows are to be controlled by a two-phase system and it is first necessary to determine the maximum value of the ratio of flow to saturation flow for each phase.

For the north/south phase the maximum y value will be denoted by $y_{max\,N/S}$. For the west/east phase, however, it is necessary to determine the greater of

$$y_w \qquad \text{or} \qquad y_{w_r} + y_e$$

Whichever is the greater will be the $y_{max\,E/W}$.

The reason for this combination of y values is that during the west/east phase the flow w continues for the whole of the green time, while flows w_r and e share this green time.

If $y_{w_r} + y_e$ is greater than y_w, then the y value for the first stage will be y_e and that for the second stage y_{w_r}. If however y_w is the greater, then the green time that flow w requires should be divided in proportion to the y values of streams w_r and e. Then the y value for the first stage is

$$\frac{y_e y_w}{y_e + y_{w_r}}$$

and for the second stage

$$\frac{y_{w_r} y_w}{y_e + y_{w_r}}$$

The same procedure may be used with a late start for the opposing flow.

Problem

The following hourly flows (table 38.1) and saturation flows relate to an intersection to be controlled by two-phase signals incorporating a late-start feature. Minimum intergreen periods are employed and starting delays are 2 s, for each green plus amber period.

Is the period that right-turning vehicles from the west approach require to complete their turning movement without obstruction from the straight ahead flow on east approach 5 s; 11 s; or 16 s?

TABLE 38.1

Approach	Flow (p.c.u./h)	Saturation flow (p.c.u./h)
west, straight ahead and left turning	400	1900
west, right turning	200	1600
east, all movements	700	1900
north, all movements	500	1900
south, all movements	600	1900

Solution

To calculate the optimum cycle time for the intersection it is first necessary to calculate the y values for the approaches. They are tabulated in table 38.2.

TABLE 38.2

	Approach	Flow (p.c.u./h)	Saturation flow (p.c.u./h)	y value
1	west, straight ahead, left turning	400	1900	0.21
2	west, right turning	200	1600	0.13
3	east, all movements	700	1900	0.37
4	north, all movements	500	1900	0.26
5	south, all movements	600	1900	0.32

$y_{max\ west/east}$ is 0.21 or 0.13 + 0.37, whichever is the greater (see chapter 38).

$y_{max\ north/south}$ is 0.26 or 0.32, whichever is the greater.

The total lost time is equal to the sum of the lost time in the intergreen period, which is 1 s (see page 274) plus starting delays of 2 s (see chapter 35) multiplied by the number of phases, that is $2(1 + 2) = 6$ s.

The optimum cycle time C_0 is given by (see chapter 36),

$$C_0 = \frac{1.5L + 5}{1 - Y}$$

$$= \frac{1.5 \times 6 + 5}{1 - (0.50 + 0.32)}$$

$$= 78 \text{ s}$$

The effective green time available for distribution between the phases is the cycle time minus the total lost time per cycle (see chapter 35).

$$= 78 - 6$$

$$= 72 \text{ s}$$

This must be divided between the phases in the ratio of the y_{max} for each phase, that is 0.50 and 0.32 (see chapter 37), giving effective green times of 44 s and 28 s respectively.

The actual green time may be obtained from the effective green times by the relationship, actual green time = effective green time + starting delays − amber period (see chapter 35).

The actual green times are thus:

      north/south phase   27 s

      east/west phase   43 s

The east/west phase is divided into two stages in the proportion of 0.13 to 0.37 giving a late start to green on the east approach of 11 s. The timing diagram is shown in figure 38.2.

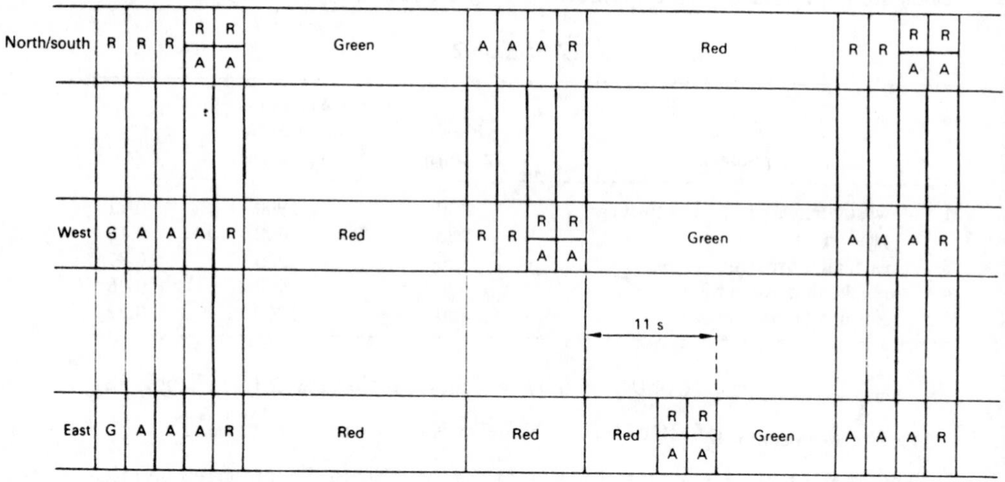

Figure 38.2

39

Opposed right-turning vehicles and gap acceptance

Right-turning vehicles present particular problems at signal-controlled intersections. On a signal approach, right-turn vehicles may be given an exclusive right-turn lane or they may be mixed with straight ahead vehicles. If vehicles do not have to give way, that is, the flow is unopposed, then the maximum discharge or saturation flow can be calculated using the relationships given in chapter 33. When the flow is opposed and the right-turn vehicles are mixed with straight ahead vehicles then once again the relationships given in chapter 33 can be used to calculate the saturation flow. If however the right-turn flow is opposed and is not mixed with straight ahead vehicles then the following simplified approach to the calculation of the right-turn discharge can be used. In this case some right-turn vehicles are able to turn through gaps in the opposing straight ahead flow and the remainder must turn in an early cut-off intergreen period at the end of the green period.

At the beginning of the green period the opposing flow is discharged at the saturation rate but later in the green period when the queue has been discharged, opposing vehicles are discharged across the stop line in the random or unsaturated manner in which they arrive on the approach.

If the length of the initial saturated green time can be calculated then the length of the unsaturated green time is also known. The discharge of right-turning vehicles through gaps in the opposing flow can then be determined from the analytical work carried out by Tanner (see chapter 18) on the discharge of vehicles at priority intersections.

This unsaturated green period during which right-turning vehicles can be discharged through the opposing flow can be determined by assuming that no vehicles remain in the opposing queue at the end of the green period. Such an assumption is correct in the flow conditions being considered.

When the saturated green time for the opposing flow is denoted by g_s, then the number of vehicles discharged during this time is

$$(r_e + g_s)q$$

where r_e is the effective red period, that is the cycle time minus the effective green period,

q is the opposing flow.

These vehicles discharge during the saturated green time g_s at the saturation flow s. Then

$$g_s = \frac{(r_e + g_s)q}{s}$$

giving

$$g_s = \frac{r_e \times q}{s - q}$$

$$= \frac{(c - g)q}{s - q} \qquad (39.1)$$

where c is the cycle time,
g is the effective green time
Let g_u be the unsaturated green time. Then

$$g_u = g - g_s$$

Substituting from equation 39.1

$$g_u = \frac{gs - qc}{s - q}$$

The number of vehicles turning right during this period n_r is then

$$n_r = s_r \left(\frac{gs - qc}{s - q} \right) \qquad (39.2)$$

where s_r is the maximum theoretical right-turning flow passing through gaps in the opposing flow.

Tanner has given the maximum intersection flow s_r as

$$\frac{q_0(1 - B_0 q_0)}{\exp[q_0(\alpha - B_0)] [1 - \exp(-B_r q_0)]}$$

where q_0 is the rate of arrival of the opposing flow in vehicles per second;
B_0 is the mean minimum time interval between opposing vehicles passing through the intersection;
B_r is the mean minimum time interval between right-turning vehicles passing through the intersection;
α is the average gap accepted by right-turning vehicles in the opposing flow.

Plotted values of the right-turning flow for values of α, B_r and B_0 typical of those noted when the opposing flow is in one or two lanes are shown in figure 39.1, which is reproduced from *Road Research Technical Paper 56*.

The difference between the average number of right-turning vehicles arriving per cycle and n_r calculated from equation 39..2 gives the number of vehicles that must turn right during an early cut-off or intergreen period.

Right-turning vehicles are assumed to discharge at headways of 2.5 s past a point in the centre of the intersection, the first right-turning vehicle passing this point at

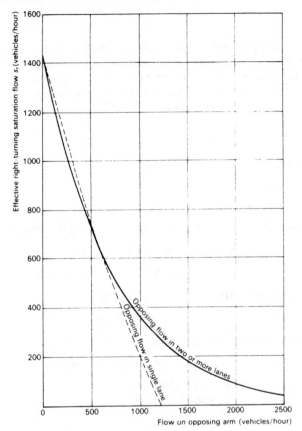

Figure 39.1   Right-turning saturation flows (based on *Road Research Technical Paper* 56)

	α	B_r	B_o
		(seconds)	
Single-lane opposing flow	6	2.5	3
Two (or more)-lane opposing flow	6	2.5	1

α is the minimum gap required in the opposing flow for 1 right-turner
B_r is the minimum headway between successive right-turners ($1/\beta_r$ is the saturation flow)
B_o is the minimum headway between successive vehicles in the opposing flow

the end of the amber period and subsequent vehicles passing this point at intervals of 2.5 s.

If it is further assumed that the first vehicle on the cross phase needs 3 s from the start of the green indication on the cross phase to reach the mid-point of the intersection then the length of the intergreen period will be 2.5 × the number of vehicles waiting to turn right.

The same procedure may be used to estimate the duration of any early cut-off period.

Problem

At an intersection with traffic flows as given in chapter 38 the effective green time on the west/east approach is 35 s, and the cycle time is 78 s. The right-turning saturation flow may be estimated from figure 39.1 for single-lane flow. The p.c.u./ vehicle ratio is 1.2. Is the average number of vehicles waiting to turn right at the end of the green period approximately 2, 5 or 8?

Solution

Using equation 39.2

$$n_r = s_r \left(\frac{gs - qc}{s - q} \right)$$

Details of the traffic flows as given in chapter 38 and in this chapter are:

q (flow on the opposing east approach) = 700 p.c.u./h

or 700/1.2 = 583 vehicles/h

g (effective green time) = 35 s

c (cycle time) = 78 s

s (saturation flow on east approach) = 1900 p.c.u./h

or 1900/1.2 = 1583 vehicles/h

From figure 39.1, when q = 583 vehicles/h
s_r = 650 vehicles/h

Then

$$n_r = \frac{650}{3600} \left(\frac{35 \times 1583 - 583 \times 78}{1583 - 583} \right)$$

$$= 1.8$$

40

The ultimate capacity of the whole intersection

It was explained in chapter 35 that the capacity of an approach is dependent on the lost time during the cycle. When the whole intersection is considered the capacity is also dependent on the total lost time on all the phases because the remainder of the time is shared equally between the phases and used as running time.

When the cycle time is calculated as described in chapter 36 and the green times apportioned as described in chapter 37, the approaches on each phase that have the highest ratio of flow to saturation flow (y value) will all reach capacity simultaneouly as traffic growth takes place.

The ultimate capacity of the intersection is then the maximum flow that can pass through the intersection with the same relative flows and turning movements on the approaches.

As traffic growth takes place the capacity of the intersection can be increased by increasing the cycle time, because the ratio of lost time to cycle time decreases. There is however a practical limit beyond which there is little gain in efficiency as the cycle time is increased and at which drivers may become impatient at not receiving a green indication for the approach at which they are queueing. This practical limit is set at 120 s and the ultimate capacity of the intersection could be calculated as the flow that could just pass through the intersection when the signals were set at this cycle time.

This capacity is however the maximum capacity and is associated with long delays. It is more usual in traffic engineering work to use a practical capacity of 90 per cent of the maximum and this will result in shorter delays.

When traffic flows are uniform, the cycle time C_m, which is just long enough to pass all the traffic that arrives in one cycle, is easily calculated. Traffic flow is however semi-random or random and time is wasted because of variability of arrival times so that the minimum cycle time C_m is associated with extremely long delays.

The minimum cycle time is given by

$$C_m = L + \frac{q_1}{s_1} C_m + \frac{q_2}{s_2} C_m + \frac{q_3}{s_3} C_m + \ldots + \frac{q_n}{s_n} C_m$$

where q_n/s_n is the highest ratio of flow to saturation flow for phase n. Then

$$C_m = L + y_1 C_m + y_2 C_m + \ldots + y_n C_m$$
$$= L + C_m \times Y$$

or

$$Y = 1 - \frac{L}{C_m}$$

The maximum possible ratio of flow to saturation flow Y that can be accommodated over all the phases when C_m is at the practical limit and Y practical is employed is

$$Y_{\text{pract.}} = 0.9 \left(1 - \frac{L}{120} \right)$$

where $Y_{\text{pract.}}$ is 90 per cent of Y, or

$$Y_{\text{pract.}} = 0.9 - 0.0075L \tag{40.1}$$

As the Y value is the sum of the maximum ratios of flows divided by saturation flows for each phase and as the saturation flow is fixed, then as the flow increases Y increases from any given present day value of Y existing, to a future maximum value of Y practical. The reserve capacity at the existing flow is thus

$$\frac{100\,(Y_{\text{pract.}} - Y_{\text{exist.}})}{Y_{\text{exist.}}} \text{ per cent} \tag{40.2}$$

Problem

For the intersection with the flows given in chapter 38 and controlled by a two-phase system incorporating a late-start feature on the east approach, is the reserve capacity of the intersection 4 per cent, 14 per cent or 24 per cent?

Solution

The approaches and their y values are reproduced from chapter 38 in table 40.1.

TABLE 40.1

	Approach	Y value
1	west, straight ahead and left turning	0.21
2	west, right turning	0.13
3	east, all movements	0.37
4	north, all movements	0.26
5	south, all movements	0.32

The y values selected as y_{max} values are

east/west phase $0.13 + 0.37 = 0.50$
north/south phase $= 0.32$

From equation 40.1

$$Y_{pract.} = 0.9 - 0.0075L$$

and $L = 6$ s (see page 297)

$$Y_{pract.} = 0.9 - 0.0075 \times 6$$

$$= 0.85$$

From equation 40.2

$$\text{reserve capacity} = \frac{100(0.85 - 0.82)}{0.82}$$

$$= 4 \text{ per cent}$$

41

The optimisation of signal-approach dimensions

The procedure previously outlined have optimised cycle time and green times so as to produce minimum overall delay for an intersection where the physical dimensions of the approach highways were fixed.

When an intersection is being re-designed or a new intersection being constructed, then it is often desirable to modify the widths of the signal approaches. The necessity for this is not difficult to see, for if the intersection is to have the capacity of the approach highways, then it must have approximately twice the width of the highway if it has a green time of approximately half the real time.

Once the approach widths have been determined, then the optimum cycle time and the green times can be calculated for minimum delay conditions. Variations in approach width will result in differing optimum cycle times and thus there is theoretically an infinite number of variations possible. In practice the choice is less because the designer must normally produce a design with a standard lane width or multiples of this width.

One approach to this problem is that made by Webster and Newby[1]. They assumed that the maximum possible rate of flow across the stop line was proportional to the widths of the approach, w_1 and w_2, as shown in figure 41.1, and also that the widened sections of the approaches had lengths d_1 and d_2, which were just long enough to accommodate the queues that could pass through the intersection during fully saturated green periods.

For the intersection shown in figure 41.1 operating under two-phase control, q_1 and q_2 are the design flows of the first and second phase the approach widths of which are w_1 and w_2 and the effective green times of which are g_1 and g_2. Then

$$cq_1 = sw_1 g_1$$

and

$$cq_2 = sw_2 g_2$$

where s is the saturation flow per unit width of road and c is the cycle time.

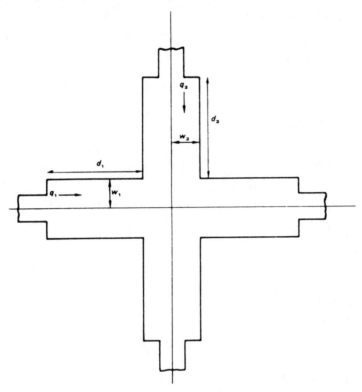

Figure 41.1   Idealised geometric layout for a signal-controlled junction with widened approaches

Since the sum of the effective green periods is constant

$$g_1 + g_2 = cq_1/sw_1 + cq_2/sw_2$$

Differentiating with respect to w gives

$$\frac{dw_2}{dw_1} = \frac{-q_1}{q_2(w_1/w_2)^2}$$

It is required to minimise the total width W ($W = w_1 + w_2$) at the intersection. For minimum total width

$$\frac{dW}{dw_1} = 0$$

or

$$1 + \frac{dw_2}{dw_1} = 0$$

Substituting for dw_2/dw_1 gives

$$1 - \frac{q_1}{q_2(w_2/w_1)^2} = 0$$

or

$$\frac{w_1}{w_2} = \sqrt{\frac{q_1}{q_2}}$$

and

$$\frac{q_1}{q_2} = \frac{w_1 g_1}{w_2 g_2} = \frac{g_1}{g_2} \sqrt{\frac{q_1}{q_2}}$$

or

$$\frac{g_1}{g_2} = \sqrt{\frac{q_1}{q_2}}$$

Also

$$\frac{d_1}{d_2} = \frac{g_1}{g_2} = \sqrt{\frac{q_1}{q_2}}$$

these rules can be summarised as

$$\frac{d_1}{d_2} = \frac{g_1}{g_2} = \sqrt{\frac{q_1}{q_2}} \tag{41.1}$$

This means that a major road carrying four times as much traffic as its minor crossroad should have approaches which are twice as wide as the minor road approaches and have green times which are twice as long.

If a width obtained from this expression results in a minor road theoretically having a width less than the practical minimum carriageway width then the practical minimum should be employed and a green time less than that theoretically necessary used. The green time saved can then be allocated to the other phases so allowing a reduction in their width.

It is often found that the flows on the differing arms of the same phase are approximately equal even though they may occur at differing times of the day. If this is not so then the highest flow should be used in the formula to obtain the width and the green time. Next the width of the approach on the same phase with the lower flow can be obtained.

The same rule can be applied to multi-phase intersections, when

$$w_1 : w_2 \ldots w_n$$

$$g_1 : g_2 \ldots g_n$$

$$d_1 : d_2 \ldots d_n$$

are proportional to

$$\sqrt{q_1} : \sqrt{q_2} \ldots \sqrt{q_n}$$

With T-junctions with 2-phase control the ratios of widths, green times and widened lengths should be

$$\frac{w_1}{w_2} = \sqrt{\frac{q_1}{2q_2}}$$

and

$$\frac{g_1}{g_2} = \frac{d_1}{d_2} = \sqrt{\frac{2q_1}{q_2}} \tag{41.2}$$

where q_1 etc., refers to the major road,
$\quad$ q_2 etc., refers to the minor road.

Reference

1. F. V. Webster and R. F. Newby, Research into the relative merits of roundabouts and traffic-signal intersections, *J. Instn Civ. Engrs*, **27** (Jan. 1964), 47–76

Problem

The traffic flows at an intersection are given in table 41.1.

TABLE 41.1

Approach	Flow (p.c.u./h)
north	3600
south	3300
east	800
west	900

Show that the relative proportions of the approach widths and effective green times on the widest approaches of each phase are 2:1 when two-phase control is used and the design minimises the area of land required.

Show also that when

(a) the cycle time is 120 s;
(b) the total lost time per cycle is 6 s;
(c) the intergreen period is 4 s;
(d) the minimum approach width on the east/west approach gives a saturation flow of 3300 p.c.u./h;

the actual green settings are 32 s and 80 s.

Solution

The two phases used for control purposes will be the north/south flow and the east/west flow (see chapter 37).

Select the maximum flow on each phase

north   3600 p.c.u./h
west   900 p.c.u./h

From equation 41.1

$$\frac{g_1}{g_2} = \frac{w_1}{w_2} = \sqrt{\frac{3600}{900}} = 2$$

The relative proportion of widths and effective green times for the widest approaches on each phase is thus 2:1.

When the cycle time is 120 s and the total lost time per cycle is 6 s, then the available effective green time is (see chapter 36)

$$120 - 6 = 114 \text{ s}$$

When this is divided between the phases in the ratio of 2:1 then

effective green time, north/south phase = 76 s

effective green time, east/west phase   = 38 s

For the west approach the minimum width is stated to give a saturation flow of 3300 p.c.u./h.

The maximum flow that can be passed with the signal settings as calculated is

$$\frac{\text{effective green time}}{\text{cycle time}} \times 3300 \text{ p.c.u./h} = \frac{38 \times 3300}{120}$$

$$= 1045 \text{ p.c.u./h}$$

This maximum flow is greater than the design flow on the west approach, which is given as 900 p.c.u./h. It is therefore possible to reduce the green time from the value calculated using equation 41.1. The effective green time required can be calculated from

$$\text{actual flow} = \frac{\text{saturation flow} \times \text{effective green time}}{\text{cycle time}}$$

or

$$\text{effective green time} = \frac{900 \times 120}{3300} = 33 \text{ s (approx.)}$$

The effective green time as calculated from equation 41.1 is 38 s and this leaves $38 - 33 = 5$ s available for use on the north/south phase. The total effective green time on the north/south phase is then $76 + 5 = 81$ s.

As the intergreen period is 4 s and the total lost time per cycle is 6 s then starting delays are 2 s of each green period.

actual green time = effective green time + lost time due to starting delays
— amber period

= effective green time − 1

= 80 s and 32 s

42

Optimum signal settings when saturation flow falls during the green period

On some traffic-signal approaches it cannot be assumed that saturation flow will remain constant throughout the greater part of the green period. The effect of blocked right-turning movements and of approaches that are wider at the stop line than on the remainder of the approach is to reduce the flow rate over the stop line as the green period proceeds.

In this problem it is the saturation flow just as the amber period commences that is important because it is this value that determines the variation in the number of

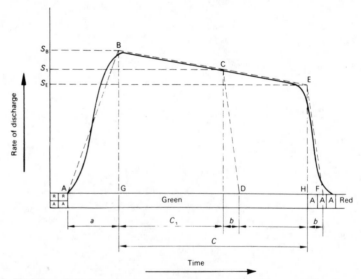

Figure 42.1   Variation of discharge across a traffic signal stop line where saturation flow decreases during the green period (based on *Road Research Technical Paper* 56)

318

vehicles crossing the stop line under saturated conditions when there is a small change in green time. It is thus necessary to find the length of green time that corresponds to the saturation flow just as amber begins and that, when substituted into the formula for optimum setting, produces the original green time.

The change in flow rate with time is illustrated for a generalised case in figure 42.1.

The rate of discharge against time curve may be replaced by a quadrilateral with the same area; this is shown by dashed lines. Then the areas ABG and EHF are equal to the areas under the corresponding portions of the curves. The saturation flow at B is S_B and at E is S_E. If the green time ends earlier at C then the saturation flow at the commencement of amber is S_1 and the reduction in flow during the amber period can be denoted by the effective line CD. From figure 42.1

$$S_1 = S_B - \frac{C_1}{C}(S_B - S_E) \tag{42.1}$$

The effective green time g_1 when the green time terminates at C can be calculated from the area of the quadrilateral ABCD, which is $\frac{1}{2}S_B a + \frac{1}{2}(S_B + S_1)C_1 + \frac{1}{2}S_1 b$. From the original definition of effective green time this area is equal to $g_1 S_1$.

Hence

$$g_1 = \frac{S_B(a + C_1) + S_1(b + C_1)}{2S_1} \tag{42.2}$$

If G is the combined green plus amber period, the lost time is

$$l_1 = G - g_1 \tag{42.3}$$

The value of g_1 is not necessarily the optimum green time. To obtain a first approximation to the optimum green time, which can then be compared with the originally assumed value of g_1, it is assumed that S_1 and l_1, calculated from equations 42.1 and 42.3, are the appropriate values of saturation flow and lost time. Then

$$C_0 = \frac{1.5L + 5}{1 - Y} \tag{42.4}$$

Also when the effective green is divided between the phases in proportion to their respective y values

$$g_1 = \frac{y_1}{Y}(C_0 - L) \tag{42.5}$$

Substituting equation 42.4 in equation 42.5

$$g_1 = \frac{y_1}{Y}\left(\frac{1.5L + 5}{1 - Y} - L\right)$$

$$= \frac{y_1}{Y(1 - Y)}(5 + L(Y + 0.5)) \tag{42.6}$$

The value of g_1 obtained from equation 42.6 is compared with the value originally obtained from equation 42.2. A new estimate of g_1 may then be made by taking a second approximation midway between the previous two values. The iteration may then be repeated until there is no significant difference between successive values of g_1.

Problem

The discharge/time curve at a traffic-signal stop line for the critical approach of phase (a) may be approximated by the curve shown in figure 42.2.

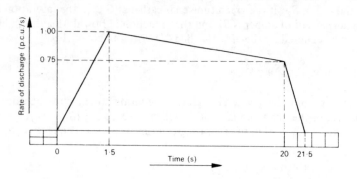

Figure 42.2

On the other approaches the saturation flow is uniform with time. The design hour flows and saturation flows for the approaches on each phase with the maximum y values are given in table 42.1.

TABLE 42.1

Approach	Design hour flow (p.c.u./h)	Saturation flow (p.c.u./h)
phase 1	600	shown graphically
phase 2	500	2,000

Intergreen period 4 s, lost time due to starting delays on phase 2 is 2 s.
    Calculate the optimum cycle time and the corresponding actual green times.

Solution

The iterative procedure to arrive at the optimum cycle time will be illustrated by carrying out the calculation in steps.

    1. Assume the actual green plus amber period on phase 1 is 18 s (point C, figure 42.1).

2. $S_1 = 1.0 - \dfrac{13.5}{18.5} (1.00 - 0.75)$

$= 0.82$ p.c.u./s (see equation 42.1)

3. $g_1 = \dfrac{1.0(1.5 + 13.5) + 0.82(1.3 + 13.5)}{2 \times 0.82}$

$= 16.55$ s   (see equation 42.2).

4. $l_1 = 18 - 16.55$

$= 1.45$ s   (see equation 42.3).

5. $L = l_1 + l_2$ + lost time in the intergreen periods

$= 1.45 + 2.00 + 1.00 + 1.00$

$= 5.45$ s

6. $y_1 = \dfrac{q_1}{S_1} = \dfrac{600}{0.82 \times 3600} = 0.30$

$y_2 = \dfrac{500}{2000} = 0.25$.

7. $g_1 = \dfrac{0.30}{0.55(1 - 0.55)} \; 5 + 5.45(0.55 + 0.5)$

$= 12.86$ s   (see equation 42.6)

8. The first calculation of actual green plus amber period on phase 1 is

12.86 s $+ 1.45$ s $= 14.31$ s   (equation 42.3)

The initial assumption of actual green plus amber period produced a value of $g_1 = 16.55$ s in Step 3 and finally in Step 7 it was calculated as 12.86 s.

A second iteration may now be performed using as the intial assumption that the actual green plus amber period is

$18 - \dfrac{16.55 - 12.86}{2} = 16.06$ s.

Successive iterations are tabulated in table 42.2.

effective green time phase 1 $= 13.52$ s

effective green time phase 2 $= \dfrac{13.52 \times 0.25}{0.30}$

$= 11.27$ s   (see chapter 37)

actual green time phase 1 $= 13.52 + 1.86 - 3.00 = 12.38$ s

actual green time phase 2 $= 11.27 - 1.00 = 10.27$ s   (see chapter 35)

cycle time $= 12.38 + 10.27 + 8.00 = 30.65$ s   (see chapter 37)

TABLE 42.2

Step	Description	Second iteration	Third iteration
1	initial value of green plus amber period	16.06 s	15.72
2	$S_1 = 1.0 - \dfrac{11.56}{18.5} (1.00 - 0.75)$	0.84 p.c.u./s	–
3	$g_1 = \dfrac{1.0(1.5 + 11.56) + 0.84(1.3 + 11.56)}{1.68}$	14.20 s	the values of the green plus amber period in the second
4	$l_1 = 16.06 - 14.20$	1.86 s	and third iterations are
5	$L - l_1 + l_2 +$ intergreen lost time	5.86 s	sufficiently close to make further iterations unnecessary
6	$y_1 = \dfrac{600}{0.84 \times 3600}$	0.30	
	$y_2 = \dfrac{500}{2000}$	0.25	
7	$g_1 = \dfrac{0.30}{0.55(1 - 0.55)} (5 + 5.86(0.55 + 0.5))$	13.52 s	

43

Delay at signal-controlled intersections

As has been briefly mentioned in chapter 36, the delay to vehicles on a traffic-signal approach has been investigated by F. V. Webster and reported in *Road Research Technical Paper* 39. By a combination of queueing theory and digital computer simulation it has been shown that the average delay per vehicle on a particular intersection arm is given by

$$d = \frac{c(1 - \lambda)^2}{2(1 - \lambda x)} + \frac{x^2}{2q(1 - x)} - 0.65\left(\frac{c}{q^2}\right)^{1/3} x^{(2 + 5\lambda)}$$

where d = average delay per vehicle,
c = cycle time,
λ = proportion of the cycle that is effectively green for the phase under consideration (that is, effective green time/cycle time),
q = flow,
s = saturation flow,
x = degree of saturation, which is the ratio of actual flow to the maximum flow that can be passed through the approach (that is $q/\lambda s$).

The relative value of the three terms is illustrated in figure 43.1 reproduced from *Road Research Technical Paper* 39. The first term in this expression is the delay due to a uniform rate of vehicle arrival, the second term is the delay due to the random nature of the vehicle arrivals. The third term was empirically derived from the simulation of traffic flow.

To assist in the calculation of delays on traffic signal approaches values of

$$\frac{(1 - \lambda)^2}{2(1 - \lambda x)} = A$$

$$\frac{x^2}{2(1 - x)} = B$$

correction term = C

are given in *Road Research Technical Paper* 39 and are tabulated in tables 43.1-3.

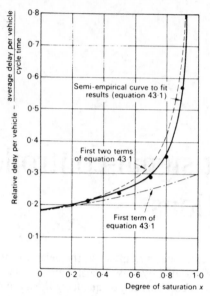

Figure 43.1 Illustration of delay on traffic-signal approaches (based on *Road Research Technical Paper 39*)

It is then possible to calculate the delay from

$$d = cA + \frac{B}{q} - C \tag{43.1}$$

TABLE 43.1 Tabulation of $A = \dfrac{(1 - \lambda)^2}{2(1 - \lambda x)}$

a/λ	0.1	0.2	0.3	0.35	0.40	0.45	0.50	0.55	0.60	0.65	0.70	0.80	0.90
0.1	0.409	0.327	0.253	0.219	0.188	0.158	0.132	0.107	0.085	0.066	0.048	0.022	0.005
0.2	0.413	0.333	0.261	0.227	0.196	0.166	0.139	0.114	0.091	0.070	0.052	0.024	0.006
0.3	0.418	0.340	0.269	0.236	0.205	0.175	0.147	0.121	0.098	0.076	0.057	0.026	0.007
0.4	0.422	0.348	0.278	0.246	0.214	0.184	0.156	0.130	0.105	0.083	0.063	0.029	0.008
0.5	0.426	0.356	0.288	0.256	0.225	0.195	0.167	0.140	0.114	0.091	0.069	0.033	0.009
0.55	0.429	0.360	0.293	0.262	0.231	0.201	0.172	0.145	0.119	0.095	0.073	0.036	0.010
0.60	0.431	0.364	0.299	0.267	0.237	0.207	0.179	0.151	0.125	0.100	0.078	0.038	0.011
0.65	0.433	0.368	0.304	0.273	0.243	0.214	0.185	0.158	0.131	0.106	0.083	0.042	0.012
0.70	0.435	0.372	0.310	0.280	0.250	0.221	0.192	0.165	0.138	0.112	0.088	0.045	0.014
0.75	0.438	0.376	0.316	0.286	0.257	0.228	0.200	0.172	0.145	0.120	0.095	0.050	0.015
0.80	0.440	0.381	0.322	0.293	0.265	0.236	0.208	0.181	0.154	0.128	0.102	0.056	0.018
0.85	0.443	0.386	0.329	0.301	0.273	0.245	0.217	0.190	0.163	0.137	0.111	0.063	0.021
0.90	0.445	0.390	0.336	0.308	0.281	0.254	0.227	0.200	0.174	0.148	0.122	0.071	0.026
0.92	0.446	0.392	0.338	0.312	0.285	0.258	0.231	0.205	0.179	0.152	0.126	0.076	0.029
0.94	0.447	0.394	0.341	0.315	0.288	0.262	0.236	0.210	0.183	0.157	0.132	0.081	0.032
0.96	0.448	0.396	0.344	0.318	0.292	0.266	0.240	0.215	0.189	0.163	0.137	0.086	0.037
0.98	0.449	0.398	0.347	0.322	0.296	0.271	0.245	0.220	0.194	0.169	0.143	0.093	0.042

TABLE 43.2   Tabulation of $B = \dfrac{x^2}{2(1-x)}$

x	0.00	0.01	0.02	0.03	0.04	0.05	0.06	0.07	0.08	0.09
0.1	0.006	0.007	0.008	0.010	0.011	0.013	0.015	0.017	0.020	0.022
0.2	0.025	0.028	0.031	0.034	0.038	0.042	0.046	0.050	0.054	0.059
0.3	0.064	0.070	0.075	0.081	0.088	0.094	0.101	0.109	0.116	0.125
0.4	0.133	0.142	0.152	0.162	0.173	0.184	0.196	0.208	0.222	0.235
0.5	0.250	0.265	0.282	0.299	0.317	0.336	0.356	0.378	0.400	0.425
0.6	0.450	0.477	0.506	0.536	0.569	0.604	0.641	0.680	0.723	0.768
0.7	0.817	0.869	0.926	0.987	1.05	1.13	1.20	1.29	1.38	1.49
0.8	1.60	1.73	1.87	2.03	2.21	2.41	2.64	2.91	3.23	3.60
0.9	4.05	4.60	5.28	6.18	7.36	9.03	11.5	15.7	24.0	49.0

TABLE 43.3   Correction term of equation 43.1 as a percentage of the first two terms

x	λ/M^*	2.5	5	10	20	40
0.3	0.2	2	2	1	1	0
	0.4	2	1	1	0	0
	0.6	0	0	0	0	0
	0.8	0	0	0	0	0
0.4	0.2	6	4	3	2	1
	0.4	3	2	2	1	1
	0.6	2	2	1	1	0
	0.8	2	1	1	1	1
0.5	0.2	10	7	5	3	2
	0.4	6	5	4	2	1
	0.6	6	4	3	2	2
	0.8	3	4	3	3	2
0.6	0.2	14	11	8	5	3
	0.4	11	9	7	4	3
	0.6	9	8	6	5	3
	0.8	7	8	8	7	5
0.7	0.2	18	14	11	7	5
	0.4	15	13	10	7	5
	0.6	13	12	10	8	6
	0.8	11	12	13	12	10
0.8	0.2	18	17	13	10	7
	0.4	16	15	13	10	8
	0.6	15	15	14	12	9
	0.8	14	15	17	17	15
0.9	0.2	13	14	13	11	8
	0.4	12	13	13	11	9
	0.6	12	13	14	14	12
	0.8	13	13	16	17	17

TABLE 43.3 (*cont*)

	0.2	8	9	9	9	8
0.95	0.4	7	9	9	10	9
	0.6	7	9	10	11	10
	0.8	7	9	10	12	13
	0.2	8	9	10	9	8
0.975	0.4	8	9	10	10	9
	0.6	8	9	11	12	11
	0.8	8	10	12	13	14

*M is the average flow per cycle = q_c.

Problems

An intersection controlled by two-phase traffic signals has the design-hour flows and saturation flows tabulated in table 43.4.

TABLE 43.4

Approach	Flow (vehicles/h)	Actual green (s)	Saturation flow (vehicles/h)
north	2000	80	3350
south	1750	80	3350
east	660	32	2750
west	750	32	2750

Starting delays 2 s of each green plus amber period intergreen period 4 s; cycle time 120 s.

For traffic flow on the north approach, is the delay element due to the random arrival of vehicles greater than the element due to the regular arrival of vehicles?

For the traffic flow on the east approach, is the average delay per vehicle approximately 3.0 s, 60 s or 9.0 s?

Solutions

First calculate the parameters of equation 43.1. For the north approach

$c = 120$ s

λ = effective green/cycle time

$= (80 + 1)/120$

$= 0.675$

$q = 2000/3600$

$= 0.556$ vehicles/s

$s = 3350/3600$

$= 0.931$ vehicles/s

$x = q/\lambda s$

$= 0.556/(0.675 \times 0.931)$

$= 0.885$

The delay element due to the uniform rate of arrival of vehicles is given by the first term of equation 43.1, that is cA.

Tabulated values of A are given in table 43.1 and referring to this table and using the approximate values $\lambda = 0.70, x = 0.90$

$A = 0.122$

and

$d = 120 \times 0.122 = 14.64$ s

The delay element due to the random rate of arrival of vehicles is given by the second term of equation 43.1, that is B/q. Tabulated values of B are given in table 43.2. Referring to this table and using the approximate values for x of 0.89

$B = 3.60$

Then

$d = 3.60/0.556$ s

$= 6.47$ s

Note. The delay element due to the uniform arrival of vehicles (14.64 s) is greater than the delay element due to the random arrival of vehicles (6.47 s).

First calculate the parameters of equation 43.1 for the east approach

$c = 120$ s

$\lambda = (32 + 1)/120$

$= 0.275$

$q = 660/3600$

$= 0.183$ vehicle/s

$s = 2750/3600$

$= 0.764$ vehicle/s

$x = 0.183/(0.275 \times 0.764)$

$= 0.871$

$M = qc$

$= 0.183 \times 120$

$= 21.96$ vehicles/cycle

From equation 43.1

$$d = cA + \frac{B}{q} - C$$

From table 43.1, $A = 0.329$.
From table 43.2, $B = 2.91$.
From table 43.3, $C = 11$ per cent of first two terms
and

$d = 49.3$ s/vehicle

44

Determination of the optimum cycle from a consideration of delays on the approach

Minimisation of delay for a complete intersection has been investigated in Appendix 3 *Road Research Technical Paper* 39 by considering the first two terms in equation 43.1. The total delay for the intersection is given by

$$D = \Sigma \text{ (average delay per vehicle)} \times \text{flow}$$

$$D = \sum_{r-1}^{r-n} \left[\frac{c(1 - \lambda_r)^2}{2(1 - \lambda_r x_r)} + \frac{x_r^2}{2 q_r (1 - x_r)} \right] q_r$$

$$= \sum_{1}^{n} \left[\frac{c y_r s_r (1 - \lambda_r)^2}{2(1 - y_r)} + \frac{y_r^2}{2 q_r (\lambda_r - y_r)} \right]$$

where $y = q/s$ for any phase.

Differentiating with respect to cycle time

$$\frac{dD}{dc} = \sum_{1}^{n} \left[\frac{(1 - \lambda_r)^2 y_r s_r}{2(1 - y_r)} - \frac{d\lambda_r}{dc} \frac{y_r^2 (2\lambda_r - y_r)}{2\lambda_r^2 (\lambda_r - y_r)} + \frac{c y_r s_r (1 - \lambda_r)}{(1 - y_r)} \right]$$

$$= 0 \text{ for minimum delay}$$

that is

$$\sum_{1}^{n} \frac{y_r s_r (1 - \lambda_r)}{1 - y_r} \left(\frac{1 - \lambda_r}{2} - C_0 \frac{d\lambda_r}{dc} \right) - \sum_{1}^{n} \frac{y_r^2 (2\lambda_r - y_r)}{2\lambda_r^2 (\lambda_r - y_r)^2} \cdot \frac{d\lambda_r}{dc} = 0$$

If λ is made proportional to y, then

$$\lambda_r = \frac{c - L}{c} \times \frac{y_r}{Y}$$

and

$$\frac{d\lambda_r}{dc} = \frac{y_r L}{Y c^2}$$

Substituting these values gives

$$\frac{1}{Y} \sum_1^n \frac{y_r s_r}{1 - y_r} (C_0^2 (Y - y_r)^2 - L^2 y_r^2) - L \sum_1^n \frac{y_r^2 (2\lambda_r - y_r)}{\lambda_r^2 (\lambda_r - y_r)^2} = 0$$

and this further reduces to

$$\frac{1}{Y} \sum_1^n \frac{y_r s_r}{1 - y_r} [C_0^2 (Y - y_r)^2 - L^2 y_r^2][(C_0^2 - 2C_0 L + L^2)]$$

$$-LY^3 C_0^3 n \times \frac{[C_0(2 - Y) - 2L]}{[C_0(1 - Y) - L]^2} = 0 \qquad (44.1)$$

which can be further simplified by approximating the terms $L^2 y_r^2$ and L^2, which are small in comparison with the other terms.

If values of average delay are calculated using equation 43.1 then a family of curves as shown in figure 36.1 are obtained. From an inspection of these curves it can be seen that the minimum cycle is approximately equal to twice the optimum cycle, which is represented by the vertical asymptote to the curve. It is thus possible to obtain a first approximation by this relationship.

The minimum cycle is just long enough to pass the traffic that arrives in one cycle; it is associated with very high delays. The minimum cycle is equal to the lost time per cycle and the amount of time necessary to pass all the traffic through the intersection at the maximum possible rate; that is

$$C_m = L + \frac{q_1}{s_1} C_m + \frac{q_2}{s_2} C_m + \ldots + \frac{q_n}{s_n} C_m$$

$$= L + C_m(y_1 + y_2 + \ldots + y_n)$$

$$= L + C_m Y$$

that is

$$C_m = \frac{L}{1 - Y}$$

or

$$C_0 = \frac{2L}{1 - Y}$$

It is now possible to replace L by $C_0(1 - Y)/2$ in those terms which in equation 44.1 are comparatively small, that is $L^2 y_r^2$ and L^2.

The term

$$[C_0^2(Y - y_r)^2 - L^2 y_r^2]$$

becomes

$$C_0^2 \left[(Y - y_r)^2 - \frac{y_r^2}{4} (1 - Y)^2 \right]$$

and the term

$$[(C_0^2 - 2C_0 L) + L^2]$$

becomes

$$(C_0^2 - 2C_0 L) \left[1 + \frac{L^2}{C_0^2 - 2C_0 L} \right] = (C_0^2 - 2C_0 L) \left[1 + \frac{(L/C_0)^2}{(1 - 2L/C_0)} \right]$$

$$= C_0(C_0 - 2L) \frac{(1 + Y)^2}{4Y}$$

Substituting these terms in equation 44.1 gives

$$\frac{(1 + Y)^2}{16 Y^5 n} \times L(C_0 - 2L) \sum_1^n \frac{y_r s_r}{1 - y_r} [4(Y - y_r)^2 - y_r^2(1 - Y)^2]$$

$$- \frac{L^2 [C_0(2 - Y) - 2L]}{[C_0(1 - Y) - L]^2} = 0$$

Let

$$E = \frac{L(1 + Y)^2}{16 n Y^5} \sum_1^n \frac{y_r s_r}{1 - y_r} [4(Y - y_r)^2 - y_r^2(1 - Y)^2]$$

Then the equation reduces to

$$(C_0 - 2L)E - \frac{L^2[C_0(2 - Y) - 2L]}{[C_0(1 - Y) - L]^2} = 0$$

or

$$C_0^3(1 - Y)^2 - 2C_0^2 L(1 - Y)(2 - Y) + C_0 L^2 \left[5 - 4Y - \frac{(2 - Y)}{E} \right]$$

$$+ L^3 \left[\frac{2}{E} - 2 \right] = 0 \qquad (44.2)$$

C_0 has been taken to be approximately $2L/(1 - Y)$; or a more accurate value can be taken as $2L/(1 - Y)F$ where F is a factor that takes into account the flows, the saturation flows and the lost time at the intersection. This makes it possible to replace L^3 in the last term in equation 44.2 to give

$$L^2 \frac{C_0(1 - Y)}{2F}$$

It is then possible to divide throughout the equation by C_0 and solve the resulting quadratic equation. This gives

$$C_0 = \frac{2L}{1 - Y} \left\{ 1 + \frac{\sqrt{[Y^2 - Y/F + 1/E + (1/F - 1)(1 - (1 - Y)/E)]} - Y}{2} \right\}$$

(44.3)

For most purposes P is close enough to unity for equation 44.3 to be rewritten as

$$C_0 = \frac{2L}{1 - Y} \left\{ 1 + \frac{\sqrt{[Y^2 - Y + 1/E]} - Y}{2} \right\}$$ (44.4)

Also

$$C_0 = \frac{2L}{1 - Y} F$$

so that for a first approximation

$$F = \left\{ 1 + \frac{\sqrt{[Y^2 - Y + 1/E]} - Y}{2} \right\}$$ (44.5)

When evaluating C_0 the first step is to calculate F from equation 44.5. If its value is not more than 10 per cent from unity then C_0 may be calculated from

$$C_0 = \frac{2LF}{1 - Y}$$

If however F is not close to unity then equation 44.3 should be used for a more accurate calculation.

This procedure is rather complicated for practical use and a relationship has been derived between $2LF$ and an expression of the form $KL + 5$. The value of K varies with junction type and with flow ratios but the most usual value adopted in practice is $K = 1.5$ giving the well-known formula

$$C_0 = \frac{1.5L + 5}{1 - Y} \qquad \text{(see chapter 36)}$$

The evaluation of C_0 from this relationship is tedious and a procedure to simplify the calculation is given in *Road Research Technical Paper* 39.

The procedure is as follows:

1. For each phase calculate $G = (3 - Y)/2Y$ and multiply each y value by G giving Gy_r for each phase.
2. From figure 44.1 determine B_r for each phase and multiply by the appropriate value of the saturation flow s_r to give $\Sigma s_r B_r$ for all n phases.

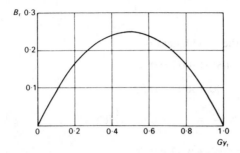

Figure 44.1 Graphical determination of B_r (based on *Road Research Technical Paper* 39)

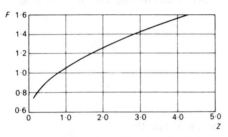

Figure 44.2 Graphical determination of F (based on *Road Research Technical Paper* 39)

3. Calculate $Z = n/L \overset{n}{\Sigma} \, s_r B_r$.
4. From figure 44.2 determine F.
5. Calculate C_0 from

$$C_0 = \frac{2LF}{1 - Y}$$

Problems

Two phase traffic-signal control is used to resolve conflicts at an intersection with the following flows and saturation flows (table 44.1). The total lost time per cycle is 6 s and the intergreen period is 4 s.

TABLE 44.1

Approach	Design-hour flow (vehicles/h)	Saturation flow (vehicles/h)
north	550	1500
south	650	1500
east	450	1500
west	500	1500

Show that the error involved in using the approximate practical formula for the optimum cycle time rather than the more accurate procedure described in *Road Research Technical Paper* 39 is less than 5 per cent.

Solutions

The y_{max} values for the two phases are

     north/south   650/1500   = 0.43

     east/west   500/1500   = 0.33

     s_r for all phases is $\dfrac{1500}{3600}$ = 0.42 vehicle/s

$$Y = 0.43 + 0.33 = 0.76$$

The approximate practical formula for the optimum cycle time is given by

$$C_0 = \frac{1.5L + 5}{1 - Y}$$

$$= \frac{1.5L + 5}{1 - 0.76}$$

$$= 58.3 \text{ s}$$

The more accurate procedure is as follows:

1. For the north/south phase

$$Gy_r = y \times \frac{3 - Y}{2Y} = 0.43 \times \frac{3 - 0.76}{2 \times 0.76}$$

$$= 0.63$$

For the east/west phase

$$Gy_r = y \times \frac{3 - Y}{2Y} = 0.33 \times \frac{3 - 0.76}{2 \times 0.76}$$

$$= 0.44$$

2. From figure 44.1

For the north/south phase and east/west phase

     $B_r = 0.24$

     $s_r B_r = 0.42 \times 0.24$

         $= 0.10$

3. $Z = n/L \sum\limits^{n} s_r B_r$

     $= 2/6(0.10 + 0.10)$

     $= 1.67$

4. From figure 44.2

     $F = 1.2$

5. $C_0 = \dfrac{2L}{1-Y} \times F$

$ = \dfrac{2 \times 6 \times 1.2}{1 - 0.76}$

$ = 60.0 \text{ s}$

The error in the approximate practical formula is then

$$\left(\dfrac{60.0 - 58.3}{60}\right) 100 \text{ per cent}$$

that is 2.8 per cent, which is less than 5.0 per cent.

45

Average queue lengths at the commencement of the green period

In the design of traffic signals it is often desirable to be able to estimate the queue length at the beginning of the green period. The queue length at this period of the cycle will normally be the greatest experienced because during the green period the queue is being discharged at a rate which for practical purposes must be greater than the flow on the approach.

Of the several cases that may be considered, the simplest one is the unsaturated approach where the queue at the beginning of the green period disappears before the signal aspect changes to amber. In this case the maximum queue is simply the number of vehicles that have arrived during the preceding effective red period, that is

$$N_u = qr \qquad\qquad (45.1)$$

where N_u is the initial queue at the beginning of an unsaturated green period,

q is the flow,

r is the length of the effective red period; the effective red period is equal to the cycle time minus the effective green period.

The next simplest case is when the approach is fully saturated. The queue length will have gradual variations over the interval considered, on which are superimposed sudden increases and decreases caused by the red and green periods. The range of the short variations will be qr and the average queue over the whole of the interval considered (assuming no great variation between beginning and end) is equal to the product of the flow and the average delay per vehicle.

The average queue at the beginning of the green period is thus equal to the average queue throughout the interval plus half the average range of the cyclic fluctuations; that is

$$N_s = qd + \tfrac{1}{2} qr$$

336

where N_s is the initial queue at the beginning of a saturated green period.

d is the average delay per vehicle on a single approach.

Note that when the approach is only just saturated then d is $qr/2$ and $N_s = qr$, the same as for an unsaturated approach. The value of N_s cannot be less than qr so that

$$N_s = qd + \tfrac{1}{2} qr \qquad \text{or} \qquad qr \tag{45.2}$$

whichever is the greater.

In most practical cases intersections will operate under a mixture of saturated and unsaturated conditions. The initial queue at the commencement of the green period for these conditions will now be discussed.

Assume there are n fully saturated cycles and m unsaturated cycles. The flow is q and the average delay per vehicle d over $(m + n)$ cycles. These are composed of flow q_s and delay d_s for n cycles and flow q_u with delay d_u for m cycles.

From the previous discussion the average queue at the beginning of the green period for the n saturated cycles will be

$$\frac{q_s r}{2} + q_s d_s$$

Similarly for the m unsaturated cycles. The average queue at the beginning of the green period for the m unsaturated cycles will be

$$q_u r$$

Therefore the average queue at the beginning of the green period over $(m + n)$ cycles will be

$$N = \frac{1}{(m + n)} \, (nq_s r/2 + nq_s d_s + mq_u r)$$

The average flow is given by

$$q = \frac{nq_s + mq_u}{(n + m)}$$

so that

$$N = qr/2 + \frac{nq_s d_s + mq_u r/2}{n + m}$$

The second expression on the right-hand side presents difficulties in evaluating a solution so it must be converted into a more easily manipulated form.

Considering the unsaturated cycles, the net rate of discharge of the queue is $s - q_u$, the average queue at the beginning of the green period is N and the time taken for this initial average queue to disperse is $N/(s - q_u)$. The average queue throughout the cycle is

$$\frac{N}{2c} \left(\frac{N}{s - q_u} + r \right)$$

as illustrated in figure 45.1.

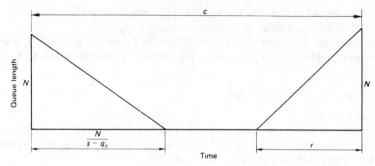

Figure 45.1   Variation of queue with time when the approach is operating under a mixture of saturated and unsaturated conditions

Average delay = average queue × cycle time/number of vehicles per cycle.

$$d_u = \text{average queue}/q_u$$

$$d_u = \frac{r}{2c}\left(\frac{rs}{c(s-q_u)}\right)$$

This is approximately $r/2$ an approximation that underestimates delay. But d_u is very much smaller than d_s and m is also much smaller than n, so that

$$N = \frac{qr}{2} + \frac{nq_s d_s + mq_u d_u}{n+m}$$

$$N = \frac{qr}{2} + qd$$

We know also that the average queue at the beginning of the green period cannot be less than qr so that the value of N is usually taken as the greater of these two values.

This theoretical approach neglects the finite extent of the queue and in practice vehicles join the queue earlier and so have an increased delay. This increase can be calculated from

length of queue when the green period begins = Nj/a

where a = number of lanes in queue,
         j = average spacing of vehicles in the queue.

Therefore

$$t = Nj/av$$

where v = free running speed of the traffic,
         t = time for a vehicle to travel the length of the queue at the running speed.
   The number of vehicles that arrive in this time is

$$\frac{qNj}{av}$$

and this correction should be added to the previously derived values of N giving

$$N = q\left(\frac{r}{2} + d\right)\left(1 + \frac{qj}{av}\right) \quad \text{or} \quad qr\left(1 + \frac{qj}{av}\right) \quad\quad (45.3)$$

whichever is the greater.

Problems

The flows and saturation flows, given in Part A below, are observed on traffic-signal approaches. Select from Part B the appropriate value of the queue at the beginning of the green period for each of the approaches described in Part A. In all cases the intergreen period is 4 s, starting delays may be taken as 2 s of each green period and the spacing of queued vehicles may be taken as 6 m. Average speed on the approach is 6 m/s.

Part A
(a) A two-lane approach with a width of 7.65 m carries a flow of 1000 p.c.u./h; the cycle time is 100 s and the actual green time 40 s. On this approach it may be assumed that vehicles arrive with a uniform headway distribution and that the saturation flow is 4015 p.c.u./h. An average vehicle is equal to 1.2 p.c.u.

(b) A single right-turning lane with a saturation flow of 1600 p.c.u./h carries a flow of 800 p.c.u./h; the actual green time is 60 s and the cycle time is 100 s. An average vehicle is equal to 1.3 p.c.u.

Part B
(a) Approximately 8 vehicles.
(b) Approximately 14 vehicles.
(c) Approximately 20 vehicles.

Solutions

(a) The saturation flow on this approach is 4015 p.c.u./h.

$525 \times w$ p.c.u./h $= 525 \times 7.65$ (see chapter 33)

$$= 4015 \text{ p.c.u./h}$$

The capacity of the approach is given by

$$\frac{\text{effective green time} \times \text{saturation flow}}{\text{cycle time}} \text{ p.c.u./h}$$

and

effective green time = actual green time + amber period − starting delays

(see chapter 37)

The capacity of the approach

$$= \frac{41 \times 4015}{100}$$

$$= 1646 \text{ p.c.u./h}$$

The flow on the approach is 1000 p.c.u./h.

The flow on the approach is less than the capacity and because the vehicles arrive at regular intervals the maximum queue at the commencement of the green period is equal to the number of vehicles arriving during the preceding red and red-plus-amber periods.

The average queue at the commencement of the green period is then

$$N_u = qr \qquad \text{(see equation 45.1)}$$

$$= \frac{1000 \times (100 - 41)}{3600} \text{ p.c.u.}$$

$$= 16.4 \text{ p.c.u.}$$

$$= 16.4/1.2 \text{ vehicles}$$

$$= 13.6 \text{ vehicles}$$

(b) The capacity of the approach $= \dfrac{61 \times 1600}{100}$ p.c.u./h

$$= 976 \text{ p.c.u./h}$$

The flow on the approach $= 800$ p.c.u./h

The approach is nearing saturation so that the average number of vehicles waiting on the approach at the commencement of the green period is given by equation 45.3

$$N = q\left(\frac{r}{2} + d\right)\left(1 + \frac{qj}{av}\right) \qquad \text{or} \qquad qr\left(1 + \frac{qj}{av}\right)$$

whichever is the greater.

The average delay on the approach d can be calculated using the method described in section 43

$$d = cA + \frac{B}{q} - C \qquad \text{(equation 43.1)}$$

and

$$\lambda = 0.61$$

$$x = 800/976 = 0.82$$

$$M = \frac{800 \times 100}{3600 \times 1.3} \text{ vehicles/cycle}$$

$$= 17.1 \text{ vehicles/cycle}$$

By interpolation from table 43.1

$A = 0.153$

By interpolation from table 43.2

$B = 1.87$

By interpolation from table 43.3

$C = 13$ per cent

$$d = 100 \times 0.153 + \frac{1.87}{0.17} - 13 \text{ per cent of previous two terms}$$

$$= 22.9 \text{ s}$$

Then

$$N = 0.17\left(\frac{39}{2} + 22.9\right)\left(1 + \frac{0.17 \times 6}{1 \times 6}\right)$$

or $0.17 \times 39\left(1 + \dfrac{0.17 \times 6}{1 \times 6}\right)$ (equation 45.3)

$$= 6.2 \quad \text{or} \quad 7.8$$

that is

$N = 7.8$ vehicles

46

The co-ordination of
traffic signals

When several traffic signal-controlled intersections occur along a major traffic route, some form of co-ordination is necessary to prevent, so far as is possible, major road vehicles stopping at every intersection. Alternatively, or in addition, the signals may be co-ordinated to minimise delays to vehicles. Sometimes linking between signals is carried out to prevent queues stretching back from one intersection to the preceding signals.

Several forms of linking between signals are possible, three of which are the simultaneous system, the alternate system and the flexible progressive system. In these systems co-ordination between intersections is usually achieved by means of a master controller.

In the simultaneous system all signals along the co-ordinated length of highway display the same aspect to the same traffic stream at the same time. Some local control is possible using vehicle actuation but a master controller keeps all the local controllers in step and imposes a common cycle time. An obvious disadvantage of this control system is that drivers are presented with several signals each with a green aspect, and there is a tendency to travel at excessive speed so as to pass as many signals as possible before they all change to red. Where turning traffic is light and intersections are closely spaced, then this system may have advantages for pedestrian movement.

The alternate system allows signal installations along a given length of road to show contrary indications. This means that if a vehicle travels the distance between intersections in half the cycle time, then a driver need not stop. The cycle time must be common to all the signals and must be related to the speed of progression. Major roads with unequal distances between intersections present difficulties for this reason.

With the flexible progressive system the green periods at adjacent intersections are offset relative to each other according to the desired speed on the highway. Progression along the highway in both directions must be considered and this usually results in a compromise between the flow in both directions and also between major and minor road flows.

A master controller keeps the local controllers, which may be either fixed-time or vehicle actuated, in step. Vehicle-actuated control is possible when there is no

342

longer a continuous demand from all detectors, the signals changing in accordance with traffic arriving at the isolated intersection.

Any change of right of way must however not interfere with the progressive plan so that at certain periods a change of right of way cannot take place because there would be insufficient time in which to regain right of way as required by the progressive plan.

Under very light traffic conditions, the flexible progressive system is likely to produce greater delays than an unlinked system because of the overriding priority of a small number of major road vehicles. In some circumstances traffic density measuring devices bring the master controller into action to impose an overall flexible progressive system as traffic volume increases, while at lower volumes the signals at each intersection act in an independent manner.

At key intersections vehicle actuation is often allowed and the control of the preceding or following signals linked to the operation of the key intersection.

Where traffic leaving the key intersection is assisted to pass through the next intersection then this is referred to as forward linking. If however traffic arriving at the key intersection on a particular approach is given preference this is referred to as backward linking. Forward linking prevents a key intersection from becoming blocked while backward linking will prevent a queue stretching back from the key intersection causing it to be blocked.

Problem

A major traffic route through a central city area has frequent regularly spaced traffic signal intersections. It is proposed to co-ordinate these signals so as to minimise delay to all vehicles passing through the intersections. Arrange the following control systems in order of their suitability for the above situation.

(a) A flexible progressive control system with fixed-time operation at each intersection.
(b) A simultaneous-control system with a major road green time of 90 s and a cycle time of 120 s.
(c) An alternate-control system with vehicle actuation at each intersection.
(d) A control system that imposes flexible progression during periods of heavier traffic flow and allows isolated-vehicle actuation at other times.

Solution

The order of merit of these control systems is:

(d) This system would allow progression during periods of heavier flow but at other times would allow the intersections to function under isolated vehicle actuated control to deal with fluctuations in the traffic flow.
(c) This system would allow progression provided the cycle time was correctly chosen and at periods of low demand vehicle actuation would allow flexibility.
(a) This system would offer progression at one level of flow but would be inflexible under changing traffic conditions.
(b) This system would not form a satisfactory method of control being unable to meet changing traffic demands and being likely to produce long delays at low traffic flows.

47

Time and distance diagrams for linked traffic signals

When the flexible progressive system of co-ordinating traffic signals is employed, then it is frequently desirable to construct a time-and-distance diagram to estimate the best offset or difference in the start of the green time, between adjacent signals.

The time and distance diagram is simply a graph on which time and hence the signal settings is plotted horizontally.

Distance along the major route between intersections is plotted vertically. The slope of any line plotted on this diagram represents the speed of progression along the major route.

Before the diagram can be prepared it is necessary to examine each of the intersections which it is desired to co-ordinate and calculate the optimum cycle times for the expected traffic flows using the relationship

$$C_{\text{pract}} = \frac{0.9L}{0.9 - Y} \quad \text{or} \quad C_0 = \frac{1.5L + 5}{1 - Y} \qquad \text{(see page 310 or 291)}$$

From these calculations it is possible to determine the intersection that is most heavily loaded and that therefore requires the longest cycle time. This intersection is then referred to as the key intersection and because in any linked system it is necessary for each cycle time to be the same, or a multiple of the same value, the cycle time of the key intersection is adopted as the cycle time for the whole system.

As traffic flows vary throughout the day and also throughout the week, it may be necessary to employ differing key intersection and common cycle times and hence differing progression plans for differing days and times.

With the cycle time c_1 for the system determined, it is possible to calculate the effective green times for the differing phases at the key intersection from the relationship

$$\text{effective } g_1 = \frac{y_1}{Y} (c_1 - L) \qquad \text{(see chapter 37)}$$

344

The actual green time may then be determined if the lost time within the green period is known (see chapter 35).

The actual green time at the key intersection gives the minimum actual green time at the other intersections for the main route with progression.

To obtain the maximum green times for the intersections other than the key intersection it is necessary to determine the shortest acceptable green times for side road phases. The shortest acceptable effective green time for a side road phase is obtained from

$$\frac{y_{side} \times c_1}{0.9}$$

and from a knowledge of the lost time in the green period the shortest acceptable actual green time can once again be calculated.

The longest actual green for the route with progression is then the linked cycle time c_1 minus the shortest acceptable actual green minus the intergreen periods.

With the calculated value of green times at the key intersection and the maximum and minimum green times at the other intersections known, it is possible to arrange the relative timing or offsets of the signals. In the preparation of this diagram the speed of progression between intersections should be chosen taking into account the known or likely speed/volume relationship for the highway and such physical features as horizontal and vertical curvature and pedestrian activity. Turning move-

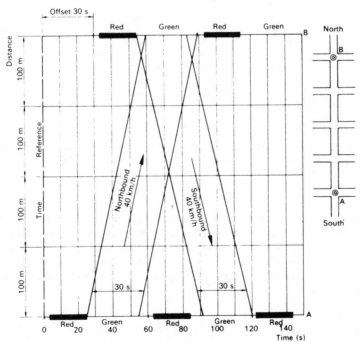

Figure 47.1   A time/distance diagram showing how a time offset is arranged between intersections A and B to give progression through two signals

ments and critical queue lengths also should be considered in the arrangement of the progression through the intersection. A typical time/distance diagram is shown in figure 47.1.

Problem

Four two-phase traffic signal-controlled intersections along a major north/south traffic route are spaced at distances of 0.5 km apart. Details of the evening peak hour traffic flows at these intersections are given in table 47.1. Starting delays in all cases may be taken as 2 s of each green period.

TABLE 47.1

Intersection	Approach	Flow	Saturation flow	Lost time per cycle (s)
A	north	1250	4015	6
	south	1450	4015	
	east	1000	2250	
	west	800	1950	
B	north	1350	4015	8
	south	1550	4015	
	east	1200	2700	
	west	650	2250	
C	north	1100	4015	8
	south	1500	4015	
	east	900	2250	
	west	550	2250	
D	north	1300	4015	8
	south	1400	4015	
	east	1000	2700	
	west	600	1950	

Prepare a time and distance diagram showing how the offsets for these signals may be arranged to produce progression for major road vehicles.

Solution

It is first necessary to calculate the optimum cycle time for each intersection, so that the maximum value can be found and adopted for the linked system (table 47.2). The longest optimum cycle time is required at the intersection B and this is the key intersection. The cycle time for the linked system is therefore 100.0 s.

The effective green time for the north/south traffic stream at intersection B is given by

$$\frac{yN/S}{Y}(c_1 - L) = \frac{0.39}{0.83}(100.0 - 8.00)$$
$$= 43.2 \text{ s}$$

and actual green time = effective green time + lost time due to starting delays
— amber period

$$= \text{effective green time} + 2 - 3$$

$$= 42.2 \text{ s}$$

This is also the minimum actual green time for the north/south traffic stream at the remaining intersections.

TABLE 47.2

Intersection	Approach	y value	C_0
A	north	0.31	
	south	0.36	$\dfrac{1.5 \times 6 + 5}{1 - (0.36 + 0.44)} = 70.0 \text{ s}$
	east	0.44	
	west	0.41	
B	north	0.34	
	south	0.39	$\dfrac{1.5 \times 8 + 5}{1 - (0.39 + 0.44)} = 100.0 \text{ s}$
	east	0.44	
	west	0.29	
C	north	0.27	
	south	0.37	$\dfrac{1.5 \times 8 + 5}{1 - (0.37 + 0.40)} = 73.9 \text{ s}$
	east	0.40	
	west	0.25	
D	north	0.32	
	south	0.35	$\dfrac{1.5 \times 8 + 5}{1 - (0.35 + 0.37)} = 60.7 \text{ s}$
	east	0.37	
	west	0.31	

The maximum actual green times at the remaining intersections are then calculated from a consideration of the minimum effective green time required for side road traffic using the relationship

$$\frac{y_{\text{side}} \times c_1}{0.9}$$

With minimum and maximum actual green times determined it is possible to plot a time and distance diagram by a trial and error process. As the intersections are regularly spaced, it is not too difficult to arrange for progression of the major road traffic streams in both directions. The diagram allows for progression, so that vehicles travel between intersections in half a cycle giving a speed of 36 km/h. The timing diagram for this very simple situation is shown in figure 47.2.

(1)	(2)	(3) Minimum effective green on minor route $\dfrac{Y_{side} \times c_1}{0.9}$	(4) Minimum actual green on minor route $(3) - 1$	(5) Maximum actual green on major route $c_1 - (40) -$ intergreen time*
Intersection	$y_{side\ max}$	(s)	(s)	(s)
A	0.44	48.9	47.9	44.1
C	0.40	44.4	43.4	46.6
D	0.37	41.1	40.1	49.9

*Intergreen time = $\dfrac{(\text{lost time} - \text{starting delays} \times 2)}{2}$ + 3 s per phase   (see chapter 31).

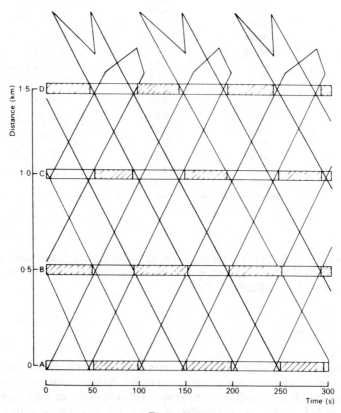

Figure 47.2

48

Platoon dispersion and the linking of traffic signals

A trial and error approach such as is involved in the preparation of a time-and-distance diagram can produce reasonable progression along a major traffic route. Where however it is desired to minimise delay at signal-controlled intersections in a network a more rigorous approach is desirable.

An alternative method of linking traffic signals so as to minimise delay is to consider the one-way stream of vehicles which after release from one traffic signal travel to the next controlled intersection and are then released when the signals change to green.

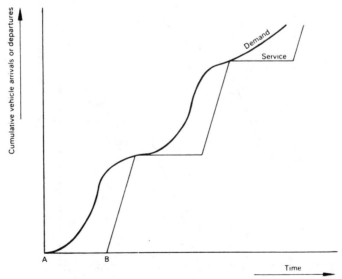

Figure 48.1 Cumulative demand and service volumes on a traffic signal approach

If the cumulative demand function, or the number of vehicles arriving at a traffic-signal approach, and the cumulative service function, or the number of vehicles discharging from a traffic-signal approach, are known then the area between the curves gives the delay on the approach. These functions are illustrated in figure 48.1.

The vertical distance between the curves represents the number of vehicles delayed at the stop line at any instant while the horizontal distance between the curves represents the duration of delay.

If the time interval between the arrival of the first vehicle at the stop line (A) and the departure of the first vehicle from the stop line (B) is varied then the area between the curves and hence the total delay is varied. By adjustment of the time interval A–B the delay may be minimised.

Such an approach can be used to minimise delay at intersections provided the forms of the demand and service functions are known. There is little difficulty in defining the service function for it is simply illustrated by variations in the flow across the stop line (see chapter 35).

It is the form of the demand function that must be defined and in a linked-signal system the problem is to predict how platoons of vehicles that are released from a traffic-signal approach disperse as the vehicles travel down the highway towards the next stop line.

Problem

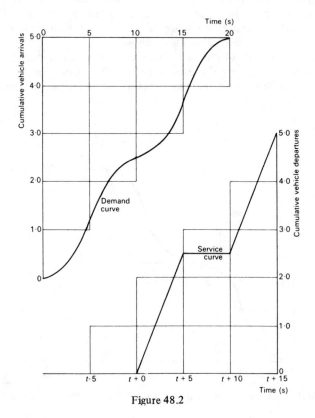

Figure 48.2

The cumulative demand function and the cumulative service functions for a traffic-signal approach are given in figure 48.2. Is the optimum time t for the commencement of green so as to minimise delay 0, 1.5, 2.5 or 4.5 s?

Solution

The offset of the green signal in relation to the arrival of the first vehicle at the stop line so as to minimise delay can be obtained by the graphical superposition of the service curve on the demand curve. In doing this it is necessary to minimise the area between the curves.

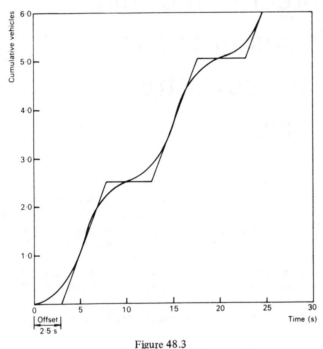

Figure 48.3

This is shown in figure 48.3 where it can be seen that for minimisation of the area between the curves the time offset should be 2.5 s.

49

The prediction of the dispersion of traffic platoons downstream of signals

The minimisation of delay by the adjustment of the offset of the green signal depends on the ability to predict the arrival rate of vehicles at the stop line. In a network controlled by traffic signals the platoons of vehicles travelling towards the stop line will have been discharged initially from a traffic signal upstream. For this reason the prediction of the dispersion of vehicles downstream from traffic-signal approaches is of considerable importance.

A considerable amount of research has been carried out on the prediction of platoon dispersion downstream of traffic signals but probably the technique that has received the greatest application is that given by D. J. Robertson[1].

Platoon dispersion is easy to predict using the following recurrence relationship

$$q_2(i + t) = Fq_1(i) + (1 - F)q_2(i + t - 1) \qquad (49.1)$$

where $q_2(i)$ is the derived flow in the ith time interval of the predicted platoon at a point along the road,
$q_1(i)$ is the flow in the ith time interval of the initial platoon at the stop line;
t is 0.8 times the average journey time over the distance for which the platoon dispersion is being calculated (measured in the same time intervals used for $q_1(i)$ and $q_2(i)$, which is usually 1/50th of the cycle time),
F is a smoothing factor.

Using data from four highway sites in London an expression for F has been given as

$$F = 1/(1 + 0.5t) \qquad (49.2)$$

Reference

1. D. J. Robertson, Transyt: a traffic network study tool, *Road Research Laboratory Report* 253 (1969)

Problems

A highway link commences and terminates with traffic signal-controlled intersections. The discharge from the first intersection may be assumed to have a uniform rate of 1 vehicle/unit time and commences at zero time. The green time at the first inter-section is 5 time units and at the second intersection 8 time units. The average travel time of the platoon between intersections may be taken as 10 time units and the travel time of the queue leader as 8 time units.

Using the recurrence relationship derived by Robertson is the percentage of vehicles delayed at the end of the green period, when the offset of the second signal from the first signal is 8 time units, approximately 25 per cent, 27.5 per cent, 35 per cent?

Solutions

The average journey time is given as 10 time units. Then

$$t = 0.8 \times \text{average journey time}$$

$$= 8 \text{ time units}$$

and

$$F = 1/(1 + 0.5t) \qquad \text{(see equation 49.2)}$$

$$= 0.2$$

Using the recurrence relationship given in equation 49.1

$$q_2(i + t) = Fq_1(i) + (1 - F)q_2(i + t - 1)$$

TABLE 49.1

				Cumulative	
i	$q_1(i)$	$i + t$	$q_2(i + t)$	Demand	Service
1	1.0	9	$0.2 \times 1.0 + 0.8 \times 0 = 0.20$	0.20	1.0
2	1.0	10	$0.2 \times 1.0 + 0.8 \times 0.2 = 0.36$	0.56	2.0
3	1.0	11	$0.2 \times 1.0 + 0.8 \times 0.36 = 0.49$	1.05	3.0
4	1.0	12	$0.2 \times 1.0 + 0.8 \times 0.49 = 0.59$	1.64	4.0
5	1.0	13	$0.2 \times 1.0 + 0.8 \times 0.59 = 0.67$	2.31	5.0
6		14	$0.2 \times 0 + 0.8 \times 0.67 = 0.54$	2.85	6.0
7		15	$0.2 \times 0 + 0.8 \times 0.54 = 0.43$	3.28	7.0
8		16	$0.2 \times 0 + 0.8 \times 0.43 = 0.34$	3.62	8.0

Total platoon content = initial discharge × time

$$= 1.0 \text{ vehicles/time unit} \times 5 \text{ time units}$$

$$= 5.0 \text{ vehicles}$$

Cumulative demand during green period at the second intersection is 3.62 vehicles.

Number of vehicles delayed = 5.0 − 3.62 = 1.38 vehicles

$$= 27.5 \text{ per cent (approximately)}$$

50

The delay/offset relationship and the linking of signals

By the use of the technique of calculating the delay to vehicles on a traffic-signal approach from the difference between the demand and service distributions, as discussed in chapter 48, and the prediction of the demand distribution from a knowledge of platoon diffusion relationships as described in chapter 49, it is possible to obtain a relationship between the offset of one signal relative to another and the delay to vehicles passing through both signals.

Using this approach it is possible to calculate the offset or difference in time between the commencement of green for two linked signals likely to result in minimum delay. The process can be extended using a method proposed by the Transport and Road Research Laboratory to a highway network. In this way a series of signal offsets can be obtained for different sets of traffic flow conditions. If signal indications are changed by a central controller or computer then when changes in traffic flow take place it will be possible to select the appropriate offset sequence to minimise delays to vehicles travelling through the network.

A method of using delay/offset relationships to obtain the relative timings or offsets of traffic signals in a network has been proposed by P. D. Whiting of the Transport and Road Research Laboratory and is referred to as the Combination method.

The basic assumptions made in the Combination method are:

(i)   The settings of the signals do not affect the amount of traffic or the route used.
(ii) All signals have a common cycle (or have a cycle that is a submultiple of some master cycle).
(iii) At each signal the distribution of the available green time among the phases is known.
(iv) The delay to traffic in one direction along any link of the network depends solely on the difference between the settings of the signals at each end of the link; it is not affected by any other signals in the network.

355

(The term 'link' is used to refer to the length of road between signalised inter-sections, because this is the more familiar usage when discussing networks. Thus 'link' and 'section' are broadly synonymous.) It is assumed unless otherwise stated that the following basic data relating to the network is available:

(i) the cycle time to be used for the network;
(ii) the delay/difference-of-offset relation in each direction for each link. The delay/difference-of-offset relation comprises N delays, one for each of the N possible differences of offset between the signals at each end of the link.

Delays in links can be combined in two ways:

1. In parallel.
2. In series.

Consider two one-way links in parallel between signalised intersections as illustrated in figure 50.1 for which the delay/offset relations are also given.

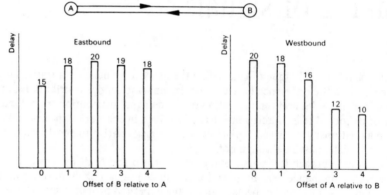

Figure 50.1   Delay/oofset relationships for link AB

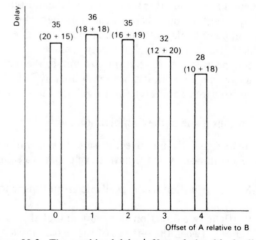

Figure 50.2   The combined delay/offset relationship for link AB

The combined delay/difference of offset relationship for the link can then be obtained by addition, taking care to add delays so that they are relative to the same signal. The total relationship is shown in figure 50.2.

It can be seen from this combined relationship that minimum delay occurs when the offset of A relative to B is 4.

When links are arranged in series, the delay difference in offset relationship for each difference of offset between the extremities has to be calculated and the minimum value selected. Consider the two links in series AB and BC shown in figure 50.3 for which the delay/offset relationships are also given.

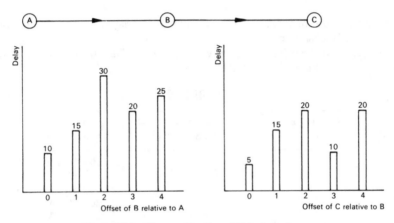

Figure 50.3   The combination of links in series

Consider initially the offset of C relative to A to be 0. Then the combined delay/ offset relationship is shown in table 50.1.

TABLE 50.1

Offset of B relative to A	Offset of C relative to B	Total delay
0	0	10 + 5 = 15*
1	4	15 + 20 = 35
2	3	30 + 10 = 40
3	2	20 + 20 = 40
4	1	25 + 15 = 40

When the offset of C relative to A is 0 minimum delay (marked *) occurs when the offset of B relative to A is 0.

Consider next the offset of C relative to A to be 1. Then the combined delay/ offset relationship is as shown in table 50.2.

TRAFFIC SIGNAL CONTROL

TABLE 50.2

Offset of B relative to A	Offset of C relative to B	Total delay
0	1	10 + 15 = 25
1	0	15 + 5 = 20*
2	4	30 + 20 = 50
3	3	20 + 10 = 30
4	2	25 + 20 = 45

When the offset of C relative to A is 1 minimum delay (marked *) occurs when the offset of B relative to A is 1.

Consider next the offset of C relative to A to be 2. Then the combined delay/offset relationship is as shown in table 50.3.

TABLE 50.3

Offset of B relative to A	Offset of C relative to B	Total delay
0	2	10 + 20 = 30*
1	1	15 + 15 = 30*
2	0	30 + 5 = 35
3	4	20 + 20 = 40
4	3	25 + 10 = 35

When the offset of C relative to A is 2 minimum delay (marked *) occurs when the offset of B relative to A is 2.

Consider next the offset of C relative to A to be 3. Then the combined delay/offset relationship is as shown in table 50.4.

TABLE 50.4

Offset of B relative to A	Offset of C relative to B	Total delay
0	3	10 + 10 = 20*
1	2	15 + 20 = 35
2	1	30 + 15 = 45
3	0	20 + 5 = 25
4	4	25 + 20 = 45

When the offset of C relative to A is 3 minimum delay (marked *) occurs when the offset of B relative to A is 0.

Finally when the offset of C relative to A is 4, the combined delay/offset relationship is as shown in table 50.5.

TABLE 50.5

Offset of B relative to A	Offset of C relative to B	Total delay
0	4	10 + 20 = 30
1	3	15 + 10 = 25*
2	2	30 + 20 = 50
3	1	20 + 15 = 35
4	0	25 + 5 = 30

When the offset of C relative to A is 4 minimum delay (marked *) occurs when the offset of B relative to A is 1.

It is now possible to produce a combined delay/difference of offset relationship, as shown in figure 50.4.

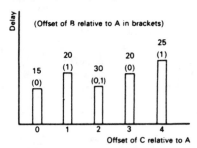

Figure 50.4   The combined delay/offset relationships for links AB and BC

It can be seen that minimum delay for the combined link occurs when the offset of C relative to A is 0 and when the offset of B relative to A is 0 respectively.

Problem

For the network and delay/offset relationships shown in figure 50.5 determine the offsets for minimum delay in the network.

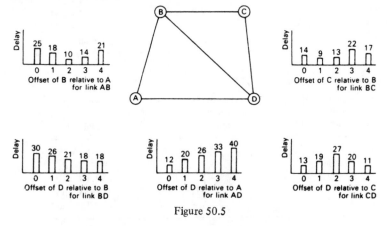

Figure 50.5

Solution

Combination of links BC and CD. (A series combination.)

1. *Difference of offset between D and B of 0*

Offset of C relative to B	Offset of D relative to C	Total delay
0	0	14 + 13 = 27
1	4	9 + 11 = 20*
2	3	13 + 20 = 33
3	2	22 + 27 = 49
4	1	17 + 19 = 36

2. *Difference of offset between D and B of 1*

Offset of C relative to B	Offset of D relative to C	Total delay
0	1	14 + 19 = 33
1	0	9 + 13 = 22*
2	4	13 + 11 = 24
3	3	22 + 20 = 42
4	2	17 + 27 = 44

3. *Difference of offset between D and B of 2*

Offset of C relative to B	Offset of D relative to C	Total delay
0	2	14 + 27 = 41
1	1	9 + 19 = 28
2	0	13 + 13 = 26*
3	4	22 + 11 = 33
4	3	17 + 20 = 37

4. *Difference of offset between D and B of 3*

Offset of C relative to B	Offset of D relative to C	Total delay
0	3	14 + 20 = 34
1	2	9 + 27 = 36
2	1	13 + 19 = 32
3	0	22 + 13 = 35
4	4	17 + 11 = 28*

5. Difference of offset between D and B of 4

Offset of C relative to B	Offset of D relative to C	Total delay
0	4	14 + 11 = 25*
1	3	9 + 20 = 29
2	2	13 + 27 = 40
3	1	22 + 19 = 41
4	0	17 + 13 = 30

The minimum delay values are marked * and for the varying differences of offset between D and B the delay histogram can be plotted when BC and CD are combined (figure 50.6).

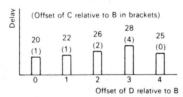

Figure 50.6   The series combination of links BC and CD

Combination of above and link BD (a parallel combination).   The delay for link BD is known for the offset of D relative to B and so the combination of the two links is a straightforward addition, giving figure 50.7.

Figure 50.7   The parallel combination of link BD and the previous combination

Combination of above and AB (a series combination)

6.   *Difference of offset between D and A of 0*

Offset of D relative to B	Offset of B relative to A	Total delay
0	0	50 + 25 = 75
1	4	48 + 21 = 69
2	3	47 + 14 = 61
3	2	46 + 10 = 56*
4	1	43 + 18 = 61

7. *Difference of offset between D and A of 1*

Offset of D relative to B	Offset of B relative to A	Total delay
0	1	50 + 18 = 68
1	0	48 + 25 = 71
2	4	47 + 21 = 68
3	3	46 + 14 = 60
4	2	41 + 10 = 53*

8. *Difference of offset between D and A of 2*

Offset of D relative to B	Offset of B relative to A	Total delay
0	2	50 + 10 = 60
1	1	48 + 18 = 66
2	0	47 + 25 = 72
3	4	46 + 21 = 67
4	3	43 + 14 = 57*

9. *Difference of offset between D and A of 3*

Offset of D relative to B	Offset of B relative to A	Total delay
0	3	50 + 14 = 64
1	2	48 + 10 = 58*
2	1	47 + 18 = 65
3	0	46 + 25 = 71
4	4	43 + 21 = 64

10. *Difference of offset between D and A of 4*

Offset of D relative to B	Offset of B relative to A	Total delay
0	4	50 + 21 = 71
1	3	48 + 14 = 62
2	2	47 + 10 = 57*
3	1	46 + 18 = 64
4	0	43 + 25 = 68

The minimum values are marked *. For varying offsets between D and A the delay histogram can be plotted (figure 50.8) when AB is added to the previous combinations.

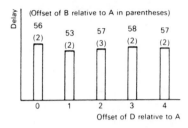

Figure 50.8   The series combination of link AB and the previous combination

Finally the previous combinations can be combined with the link AD (a parallel combination).

11. The combination is a straightforward addition giving the histogram of figure 50.9.

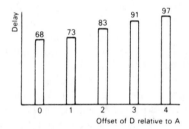

Figure 50.9   The parallel combination of link AD and the previous combination

Minimum delay for the total combination thus occurs when the offset of D relative to A is 0 (para 11).

When the offset of D relative to A is 0 minimum delay occurs when the offset of D relative to B is 3 and the offset of B relative to A is 2 (para 6).

When the difference of offset between D and B is 3 minimum delay occurs when offset of C relative to B is 4 and the offset of D relative to C is 4 (para 4).

51

Urban traffic control systems

During the late 1950s, proposals were made that the linking and co-ordination of traffic signals which had already been applied to produce green waves of traffic along major routes could be extended to a network of highways over an area. Originally referred to as the 'area control of traffic', co-ordination over a network is now more usually termed 'urban traffic control'.

Urban control strategies often seek to minimise delay over a network in contrast to the linking of signals along specified traffic routes which minimises delay to major route traffic at the expense of vehicles on the remainder of the network. The actual co-ordination of signals will depend upon the traffic and road pattern and the objective which is to be attained.

One of the first reported applications of signal control by computer was in Toronto where a pilot project was commenced in 1959; this was judged to be effective and was subsequently extended. Subsequently, the use of computers in signal control was considered for use in the United Kingdom during the 1960s where the extensive use of vehicle actuation and differences in the road pattern initially raised doubts as to the most effective application of computer control.

Two experimental systems were set up to assess the value of computer control in the United Kingdom, one was in West London and an extensive series of trials was carried out in Glasgow.

The area selected for study in West London covered an area of 6 square miles bounded by the River Thames on the south, Hyde Park on the north, Sloane Street on the east and Fulham Palace Road on the west. Within this area all the signals were already vehicle actuated and some were already linked. In particular, a linked system already operated along Cromwell Road, which at that time carried 38 000 vehicles per day towards central London.

In a description of this scheme, Mitchell[1] stated that the main intention was to establish progression over the greatest route mileage practicable whenever traffic conditions were favourable.

During the Glasgow experiment a number of systems were implemented to test their effectiveness in minimising delay. These systems could be divided into three main categories: fixed time plans based on historical data and calculated offline by a computerised optimising technique, co-ordinated systems with local traffic response at each signal, and fully responsive systems. In the first group were the Combination method, the Transyt and Sigop systems. In the second group were the Flexiprog and Equisat systems, and in the third group the systems Dynamic plan generation and Plident were tested.

Results of these experiments have been summarised by Holroyd and Robertson[2]. It was found that the Combination method, when compared with the vehicle actuated system previously operating in Glasgow, gave consistently shorter journey times averaging 12 per cent. The Transyt system gave a 4 per cent saving in journey time compared with the Combination method; the Transyt optimising program is discussed later in greater detail. The Sigop program was compared with the Transyt system during the morning and evening peaks and the off-peak period, but did not indicate any improvement over the Transyt program.

In the second group of programs, the Flexiprog program which used the Combination method and the Transyt program for basic linking together with local response to changing traffic conditions was not found to offer any improvement in journey times. Isolated vehicle actuation was also compared with the Transyt program under very light flow conditions and no measurable difference in journey times was observed. A further program in this group was the Equisat program which used the Combination method for basic linking and, once again, no measurable difference in journey times was found.

The third group of control programs, which were tested in the Transport and Road Research Laboratory, comprised dynamic plan generation and the Plident program. The former program was tested in Madrid and compared with an optimised fixed time system but produced longer journey times. The Plident program was compared with the Combination method but, once again, gave longer journey times.

As a result of these experiments the Transyt program was adopted during the early 1970s as a means of optimising traffic flow through urban networks.

The Transyt program

The Transyt program, which stands for Traffic Network Study Tool, is a digital computer program which has the object of automatically finding the best timing which will co-ordinate the operation of a network of traffic signals. After the first program was developed in 1967 a number of versions have been produced, all of which have two main elements: a traffic model and a signal optimiser.

The traffic model represents traffic behaviour in a highway network in which most junctions are controlled by traffic signals. The model predicts the value of a 'Performance Index' for the network for a given fixed time plan and an average set of flows. The Performance Index measures the overall cost of traffic congestion and is usually a combination of the total delay and the number of stops made by vehicles.

The optimisation process adjusts the signal timings and, using the traffic model, calculates if the adjusted timings reduce the Performance Index. By successive adoption of beneficial timings an optimum is reached.

Transyt assumes that all major junctions in the network are signal or priority controlled, that all the signals have a common cycle time, and that all signal stages and their minimum periods are known. For each distinct traffic stream it is assumed that, for traffic flowing between junctions, or turning at junctions, the flow rate, averaged over a specified period, is known and assumed to be constant.

In Transyt, the network is represented by nodes for the signal-controlled intersections and each distinct one-way traffic stream by links. A link may represent one or more traffic streams and one signal approach may be represented by several links. It is a matter of judgement which traffic streams warrant separate representation; two approach lanes where vehicles are equally likely to join either lane can be represented by one link whereas, if one of the lanes contains separately signalled right-turn vehicles, then a separate link is necessary.

The common cycle time necessary for optimisation is divided into a number of equal intervals or steps which are usually from 1 to 3 seconds duration. All calculations are made on the basis of the average flow rates during each step of the cycle using a histogram representing the traffic arrivals and departures at and from nodes; the dispersion of a departure stream before it arrives at the next downstream node is estimated by means of the smoothing function described in chapter 49.

Delay in the Transyt program is obtained from a consideration of these histograms or cyclic flow profiles. Two further elements of delay have to be added to this value of uniform average delay. If the arrival traffic flow exceeds capacity then the approach is saturated and an element of oversaturation delay must be added. In addition, there is an element of delay associated with the random nature of traffic flow. These latter two types of delay are calculated from simple equations.

Vehicle stops are included in the Performance Index and, as with the calculation of delay, Transyt calculates the total rate at which vehicles are forced to stop on a link as the sum of uniform, random and oversaturation flow rates. The uniform component is calculated from the cyclic flow profiles and the random and oversaturation elements from simple equations.

In most road networks there are a number of give-way junctions in addition to the signal-controlled junctions. The Transyt program allows priority junctions to be represented by referring to the major road flow as the controlling link; the flow from the non-priority link depends upon the controlling link flow, so producing a flow profile from the give-way link.

Transyt allows five separate classes of vehicles to be represented in any one queueing situation where the vehicles may in actual fact be mixed together. These classes may not be different types of vehicles but vehicles which have entered the road system from differing origins. It is thus possible to represent buses on different routes with different bus stops. This facility, referred to as a 'shared stopline', also allows the development of special 'green-wave' fixed time plans for emergency vehicles.

Fuel consumption estimates can also be output by the Transyt program for a particular set of signal timings. This estimate of fuel consumption is composed of three elements: fuel consumed while travelling at constant cruise speed between stoplines, extra fuel used during delay, and the extra fuel used in making each full stop and having to resume cruise speed.

The optimisation model in Transyt seeks the best fixed time which minimises the Performance Index, a measure of vehicle stops and delays in the network. The index is evaluated in monetary terms by assuming appropriate values for delay and stops. Optimisation of green times are also possible using Transyt but optimisation of cycle times is a more complex problem. However, Transyt has the option of automatically selecting a range of cycle times, and the cycle time with the best Performance Index output with a warning that the choice is made on the basis of unco-ordinated signal calculations.

Before Transyt can carry out the offline optimisation process, it is necessary to select the network which is to be optimised, with the links and nodes being numbered in the order in which traffic enters the road system. A decision will have to be made of the number of optimised fixed time plans which are necessary. As plan changes impose additional delay on network traffic and as good optimised plans are not very sensitive to changes in flow, the number of plan changes should be kept to a minimum and a recommended number is in the region of ten. Traffic data which must be collected is journey time and speed, saturation flows, demand flows, and turning movements.

This data is then input to the Transyt program which is run offline to produce an optimised plan. Because these optimised plans are based on past or historical records of traffic plan, the best form of control may not be achieved if there are large random variations in flow or if an accident changes the traffic flow pattern. Because of the cost and the time required to collect traffic data, it is often out of date. Nevertheless, the use of fixed time plans produced by such optimisation programs as Transyt is effective provided the required traffic engineering resources are available.

To overcome these problems experienced by fixed time optimised plans, a fully responsive urban traffic control system which does not require the offline pre-calculation has been developed by the Transport and Road Research Laboratory in collaboration with Ferranti, GEC and Plessey traffic signal system companies.

The Scoot traffic model

As a result of this co-operative effort, the Scoot (Split, Cycle and Offset Optimising Technique) has been developed and put into service. It has been stated[3] that the structure of Scoot is similar to that of the Transyt method of calculating fixed time plans. Both use a traffic model which predicts delays and stops with given signal settings, but whilst Transyt does this offline predicting delays and stops from average flows, Scoot makes these recalculations every few seconds from the latest measurements of traffic behaviour.

Both Scoot and Transyt signal optimisers automatically make systematic trial alterations to current signal timings and implement only those alterations which improve the Performance Index. In the case of Transyt, any new fixed time plan must be stored in the library of the urban traffic control computer.

The Scoot traffic model uses vehicle presence measurements obtained from road detectors which are located as far as possible from the signal stopline, ideally, just downstream of the adjacent signal-controlled junction provided a detector in this position can detect all vehicles approaching the stopline. Information from the detectors is stored in the Scoot computer in the form of cyclic flow profiles for each signal approach. The cyclic flow profile represents rate of traffic flow against time and the data is repeated with each cycle of traffic flow, hence the term 'cyclic flow profile'. Unlike the flow profiles which have to be calculated in the Transyt program, the Scoot program uses a cyclic flow profile in which the more recent data on traffic flow are combined with existing values so that large random fluctuations in the profiles are averaged out so as to be typical of the existing traffic situation.

These cyclic flow profiles are used by the signal optimiser to find the best signal timings which are the best compromise for co-ordination along all the streets in the Scoot area. As traffic flows change in the network, the cyclic flow profiles will also change and cause the signal optimiser to search for new timings.

Using information from the cyclic flow profiles and the cruise times between the detectors and the downstream stop lines, the Scoot traffic model predicts queue lengths using the green and red signal timings. As in the Transyt system, the Scoot optimiser calculates a Performance Indicator which, in this case, is the average value of the sum of the queues. In a system where every vehicle received a green indication then the sum of the queues would be zero; this is not normally possible but the optimiser seeks to adjust the signal timings so that the value is as small as possible. In addition, Scoot predicts the number of vehicle stops, the proportion of the cycle time during which vehicles are stationary over the Scoot detector, and the degree of saturation on the signal approaches are these values which are employed by the optimiser to control the cycle time and green durations.

The Scoot optimiser makes frequent small changes to the signal timings as the traffic demand changes, rather than large changes which are made infrequently with systems with several fixed time plans. It has been stated[3] that the advantages of the Scoot method are considered to be as follows.

Firstly, that there are no large, sudden changes in signal timings. In a comparison of alternative methods of changing from one timing plan to another, even the best methods caused significant increases in vehicle delay during the transition period which may last for two or more signal cycles. This means that a new plan must operate for ten to fifteen minutes to ensure that total traffic benefits exceed the initial disbenefits during the change.

Secondly, there is no need to predict average traffic behaviour for several minutes into the future. If, as discussed in the previous paragraph, a new plan must remain in operation for at least ten minutes, then the average flows for the future ten minutes must be predicted in the previous period so that the new plan can be calculated online. There are difficulties in making accurate predictions because of large random variations in traffic flow which disguise longer-term trends. The use of small frequent timing alterations allows Scoot to follow changes in traffic behaviour without requiring longer-term predictions of average flows.

Thirdly, the sensitivity of Scoot to faulty information from vehicle detectors is reduced because new signal timings result from the accumulation of a large number of small changes, hence a few poor decisions by the optimiser are of limited importance.

When Scoot adjusts green durations the 'split' optimiser estimates, a few seconds before each stage change, whether it is better to change the stage earlier, as scheduled or later. Each decision may alter the stage change by only a few seconds based on the desire to minimise the maximum degree of saturation on the junction approaches, taking into account current estimates of queue length, congestion on approaches and input minimum green times. What are referred to as 'temporary changes' are made because of cycle by cycle random variations in traffic flow, and for each temporary variation a smaller permanent variation is made to stored values of green times. The result is that green times at a junction can be completely revised to meet new traffic patterns.

Offset optimisation is achieved in the Scoot program because, once every cycle time the optimiser makes a decision whether or not to alter all the scheduled stage change times at a junction. Because all stages are changed by the same amount, it is the offset which is altered relative to other junctions. Any decisions alter the offset by a few seconds so that between adjacent junctions the offset may be altered twice every cycle. At a predetermined stage in each signal cycle, the offset optimiser considers the stored cyclic profiles and estimates whether a change in the offset will improve traffic progression on the streets immediately upstream or downstream of the junction being considered. Performance indicators on adjacent streets for offsets a few seconds earlier or later are compared and the most beneficial implemented.

The Scoot optimiser can also vary the cycle time, which is common within subareas of the signal-controlled network, by a few seconds at intervals of not less than 90 seconds. Each sub-area can be varied independently of other sub-areas with limits set by normal signal design for minimum and maximum cycle times. A decision to change the cycle time is made on the basis of maintaining, if possible, at the most heavily loaded junction in the sub-area a degree of saturation of approximately 90 per cent. This means that, in periods of low flow, the cycle times will shorten and, in periods of heavy flow, cycle times will lengthen. Less heavily trafficked junctions in a sub-area may have a low degree of saturation and, in this case, a decision to operate these signals by double cycling may take place.

Traffic surveys in Glasgow and Coventry have shown that the use of Scoot is likely to reduce average delay at traffic signals compared with good fixed time plans by about 12 per cent. It is considered that reductions in delay are likely to vary considerably from junction to junction depending on the magnitude and variability in traffic flow, the street layout and the previous standard of co-ordination between signals. The greatest benefits are likely as flows approach capacity, where flows are variable and difficult to predict and where signal-controlled junctions are relatively close.

References

1. G. Mitchell, Control concepts of the west London experiment, *Symposium on Area Control Road Traffic*, Institution of Civil Engineers, London (1967)
2. J. Holroyd and D. I. Robertson, Strategies for area traffic control systems: present and future, *Transport Road Research Laboratory Report* LR 569, Crowthorne (1973)
3. P. B. Hunt, D. I. Robertson, R. D. Brotherton and R. I. Winton, SCOOT a traffic responsive method of coordinating signals, *Transport and Road Research Laboratory Report* 1014, Crowthorne (1981)

APPENDIX

Definition of symbols used in Part 3

a = number of traffic lanes on a signal approach

A = $\dfrac{(1-\lambda)^2}{2(1-\lambda x)}$ = first term in the equation for average delay on a traffic-signal approach; due to the uniform rate of vehicle arrivals

B = $\dfrac{x^2}{2(1-x)}$ = second term in the equation for average delay on a traffic-signal approach; due to the random nature of vehicle arrivals

B_r = mean minimum time headway between right-turning vehicles as they pass through a traffic signal-controlled intersection

B_0 = mean minimum time headway between opposing vehicles as they pass through a traffic signal-controlled intersection

B_1 = mean minimum time headway between bunched vehicles on the major road

B_2 = mean minimum time headway between vehicles as they discharge without conflict from the minor to the major road

c_1 = common cycle time of linked signals

C = correction term in the equation for average delay on a traffic-signal approach

C_0 = optimum cycle time

C_m = minimum cycle time

d = average delay per vehicle on a traffic-signal approach

d_g = dummy variable for gradient

d_n = dummy variable for nearside lane

f = proportion of turning vehicles

F = smoothing factor used in the prediction of platoon dispersion

g = effective green time

g_s = saturated green time

G = percentage gradient on a signal approach

j = average spacing of vehicles in a queue on a traffic-signal approach

L = total lost time per cycle

n_ϱ = number of lanes in an opposing entry

n_r = number of right-turning vehicles

n_s = number of right-turn storage spaces within a junction without blocking following straight ahead vehicles

N = average queue at the beginning of the green period

N_s = initial queue at the beginning of a saturated green period

N_u = initial queue at the beginning of an unsaturated green period

p = conversion factor from vehicle to p.c.u.

q = flow on a traffic-signal approach

q_0 = flow on an opposing traffic-signal approach

q_1 = flow on the major road

q_2 = flow on the minor road

$q_{1(i)}$ = flow in the ith time interval of a predicted platoon at the stop line

$q_{2(i)}$ = derived flow in the ith time interval of a predicted platoon at a point along the road

r = turning radius

r_0 or r = effective red period (cycle time − effective green time)

s_0 = saturation flow of an opposing entry

S_1 = saturation flow of an individual lane

S_2 = saturation flow of an opposed lane

S_c = saturation of an opposed lane after the effective green period

S_g = saturation of an opposed lane during the effective green period

T = the through car unit of a turning vehicle

v = free running speed of vehicles on a traffic-signal approach

w = width of approach

x = degree of saturation; that is, the ratio of actual flow to the maximum flow

x_0 = degee of saturation on an opposing arm

y = ratio of flow to saturation flow = $\dfrac{q}{\lambda s}$

y_{max} = maximum value of y

Y = sum of the maximum y values for all phases comprising the cycle

Y_{pract} = 90 per cent of Y

α = average gap accepted in the opposing flow by right-turning vehicles, or the mean lag or gap in the major-road stream that is accepted by minor-road drivers

λ = ratio of effective green time to cycle time

Index